Advanced Water Technologies

Advanced Water Technologies

Concepts and Applications

P.K. Tewari

CRC Press
Taylor & Francis Group
Boca Raton London New York

CRC Press is an imprint of the
Taylor & Francis Group, an **informa** business

First edition published 2021
by CRC Press
6000 Broken Sound Parkway NW, Suite 300, Boca Raton, FL 33487-2742

and by CRC Press
2 Park Square, Milton Park, Abingdon, Oxon, OX14 4RN

© 2021 Taylor & Francis Group, LLC

First edition published by CRC Press 2021

CRC Press is an imprint of Taylor & Francis Group, LLC

Library of Congress Cataloging-in-Publication Data

A catalog record has been requested for this book

ISBN: 978-1-138-10660-4 (hbk)
ISBN: 978-1-315-10151-4 (ebk)

Typeset in Times
by SPi Global, India

Dedication

The book is dedicated to my son and daughter, Tanmay and Tanu, for their unconditional support.

Contents

Preface

This book is intended to serve as a guide to the field of advanced water treatment as well as covering conceptual aspects of water treatment and its applications. While a number of existing books address water and wastewater treatment, advanced water treatment topics including nanotechnologies, nanocomposite membranes, desalination, recycling and reuse, guidelines for setting up plant, safety considerations, and artificial intelligence (AI) and internet of things (IoT) applications are yet to be explored. Hence, there is a definite need for a book of this kind for stakeholders involved in advanced water treatment who are building capacity. The book is intended to guide professionals from chemical, civil, mechanical, AI, IoT, infrastructural and environmental engineering as well as biotechnology professionals dealing with advanced water treatment.

The water community has various stakeholders including government agencies, industries, agriculture, the service sector, research institutions, universities and colleges. Other important stakeholders are resource-rich users, resource-poor users, industrial users, technology vendors, consultants, water experts, policy makers and contractors. The book describes the technological advancements of the last few years in this field and aims to bridge gaps through capacity building, knowledge sharing and self-learning. Graduate and advanced graduate students, academic researchers and post-doctoral fellows in the field of water and wastewater treatment will find this book very useful.

Chapter 1 provides background information about water and some of its special characteristics, such as hydrogen bonding. Chapters 2–4 cover the concepts and application aspects of membrane technology, nanotechnology and nanocomposite membrane technology for water treatment. Chapter 5 deals with commercial and innovative processes for seawater and brackish water desalination. Chapters 6 and 7 discuss drinking water purification, common contaminants and treatment methodologies, as well as wastewater treatment, recycling and reuse. Chapter 8 provides guidelines for setting up plant based on advanced water technologies. Chapter 9 presents challenges and opportunities in the field of advanced water treatment. The emerging and important area of AI in water management is discussed in Chapter 10. Case studies are included in different chapters.

Chapter 1 introduces some of the special characteristics of water such as hydrogen bonding. The chapter also discusses aspects of water resource management such as water harvesting, isotope hydrology for water recharging and monitoring of contaminants.

Chapter 2 presents details of membrane technologies for water purification. The application of membrane technology as an assisted unit operation is increasing due to its superior performance, technological advances and water quality regulations that at times cannot be effectively met by conventional treatment processes. This chapter presents the preparation techniques and transport mechanism of microfiltration (MF), ultrafiltration (UF), nanofiltration (NF), reverse osmosis (RO) and electrodialysis (ED). Other membrane systems including membrane contactor, membrane distillation, biomimetic membranes, liquid membranes and ion-exchange membranes are also discussed. Different types of membrane modules such as hollow-fiber, plate-and-frame, tubular membrane and spiral-wound modules are described, together with selection criteria. Concentration polarization and membrane fouling are the major challenges in membrane separation processes. Case studies of drinking water purification using membrane processes are presented in the chapter.

Chapter 3 deals with the practical applications of nanomaterials in water purification. An overview of the different types of nanomaterials used in water treatment applications, such as metal and metal oxide nanoparticles, carbon nanotubes, zeolites and dendrimers, is given in this chapter. Different synthesis routes for nanomaterials are discussed. The environmental and health implications of the use of nanomaterials are examined.

A conventional membrane has certain inherent limitations in water treatment and purification due to the trade-off between permeability and selectivity, as well as low resistance to fouling. A nanocomposite membrane has the potential to overcome such limitations. It can be tailored to meet the requirements of specific water treatment applications by tuning structure and physico-chemical properties such as hydrophilicity, porosity, charge density, thermal and mechanical stability, as well as introducing unique functionalities. Chapter 4 presents details of different types of nanocomposite membranes: (i) conventional nanocomposite; (ii) thin-film nanocomposite (TFN); (iii) thin-film composite (TFC) with nanocomposite substrate; and (iv) surface-located nanocomposite. Challenges and opportunities for nanocomposite membranes in water purification are discussed.

Chapter 5 presents various desalination processes used to produce pure water from saline water. Multi-stage flash (MSF), multi-effect distillation (MED) and vapor compression (VC) are commercial thermal desalination processes which utilize heat energy for seawater desalination. Reverse osmosis (RO) and electrodialysis (ED) are membrane processes for water desalination. Nuclear desalination and low-carbon desalination are highlighted in connection with climate change issues. Recovery of valuable materials from brine effluent is discussed. Different aspects of brackish water desalination are presented. Alternative technologies such as solar- and wind-powered desalination, membrane distillation, capacitive deionization (CDI), freezing, water harvesting from air, submarine desalination and thermocline-driven desalination are discussed. Cost-reduction strategies achieved through technological innovation form an important part of the chapter. Case studies of seawater desalination using hybrid technology in a nuclear complex and using MED technology in a petrochemical complex are presented, as is a case study of a brackish water desalination system in a remote rural area.

Conventional water treatments such as sedimentation, coagulation, chemical disinfection and sand filtration are discussed in Chapter 6. Ceramic filtration, colloidal silver-based filtration and ion exchange are also considered. Industrial water treatment and case studies for different types of industries are presented.

Chapter 7 examines methodologies for wastewater treatment, recycling and reuse of water. Primary, secondary and tertiary treatment methods include sedimentation, biological treatment with sedimentation, and removal of residual or nonbiodegradable constituents. The membrane process helps recover a significant fraction of good-quality water for recycling. Water recovery and reuse with the potential of zero effluent discharge is gaining ground. Case studies in this chapter cover wastewater treatment, recycling and reuse in different industries such as tanneries and textiles.

Chapter 8 introduces guidelines for setting up an advanced water treatment project, including selection criteria, basic considerations, project description, contract and delivery models, financial analysis and environmental impact assessment.

In order to fine tune existing technologies and convert concepts to new technologies, it is important to enhance interdisciplinary expertise. R&D input is required to translate new eco-friendly, simple and efficient concepts utilizing new and renewable energy sources into useful technologies. Chapter 9 discusses relevant opportunities, with an emphasis on industry–academia interaction and business models. Environmental considerations and challenges for coastal and inland locations are also examined.

As an emerging technology, the application of AI and IoT has the potential to play an increasingly important role in water resource management, drinking water supply, wastewater treatment, real-time water monitoring and smart water grids. Using sensors, AI/IoT-based online and real-time monitoring of water quality and flow rate in drinking water pipeline networks can help achieve efficient operation and increase consumer confidence. Predictive real-time AI/IoT-enabled systems help evaluate the performance of water purification units. Chapter 10 highlights major AI/IoT-focused water treatment, water management and sensor development.

This book addresses a wide range of water and wastewater issues. It contains a strong focus on advanced water treatment technologies and stresses the reuse aspects of wastewater. The text is presented with clarity for the beginner, yet it is sufficiently comprehensive and thorough to be a valuable source of information for all stakeholders, which includes government agencies, industries, agriculture, the service sector, research institutions, universities and colleges. The book will be the guide of choice for the scientist and engineer, whether a student approaching the subject for the first time or a seasoned expert exploring advanced technologies. The general public, industrial users, technology vendors, consultants, policy makers and contractors will find the contents of the book enhances their understanding. This book will assist with capacity building for professionals from chemical, civil, mechanical, AI and environmental engineering as well as biotechnology professionals dealing with water treatment.

Pradip K. Tewari
Department of Chemical Engineering
Indian Institute of Technology, Jodhpur, India

About the Author

Professor Pradip K. Tewari, currently Head of the Department of Chemical Engineering at the Indian Institute of Technology (IIT), Jodhpur, is a renowned scientist specializing in membrane technology and total water solutions. Professor Tewari joined the Bhabha Atomic Research Centre (BARC) in 1977, where he was Ramanna Research Fellow, becoming a Distinguished Scientist in 2012 and Associate Director of the Chemical Engineering Group in 2014. He was a professor at Homi Bhabha National Institute (HBNI) for about ten years, and consultant to the Office of the Principal Scientific Adviser to the Government of India on water issues. Professor Tewari has been a consultant and advisor on water-related issues to several government bodies including the Department of Drinking Water and Sanitation (DDWS), Department of Science and Technology (DST), Defence Research Development Organization (DRDO), Oil and Natural Gas Commission (ONGC), Bureau of Indian Standards (BIS) and Chennai Metro Water Supply and Sewerage Board (CMWSSB). Professor Tewari received his PhD in chemical engineering from IIT Bombay in 1987. He was chairman of the International Nuclear Desalination Advisory Group and Technical Working Group on Nuclear Desalination of the International Atomic Energy Agency (IAEA) from 2005 to 2015. Dr Tewari is a Fellow of the Indian National Academy of Engineering (INAE). He is Vice President of the Asia Pacific Desalination Association (APDA). Professor Tewari has more than 200 research publications in peer-reviewed journals, proceedings, books and encyclopedias. He has two patents to his credit. His book *Nano-composite Membrane Technology* is published by CRC Press Taylor & Francis Group (USA). He is a member of the editorial board of several peer-reviewed journals such as *Desalination and Water Treatment*. He is the recipient of several individual awards and felicitations as well as group achievement awards.

1 Water

1.1 WATER—A MIRACLE COMPOUND

Water is a miracle liquid. The Earth is unique among the known celestial bodies in having three-quarters of its surface covered by water. Life has evolved in water. The regeneration and redistribution of water through evaporation and condensation is a continuous process, making it seem endlessly renewable. Water is not a commodity. It is life-making material. Whenever we look for a life, we look for a drop of water first.

The association of human beings with water starts even before birth—the fetus swims in it for nine prenatal months within the warmth of the mother's womb. The very life on this planet begins within water. Every cell within the organic structure of the human body has a fluid interior, e.g., blood is 90% water, kidneys are 82% water, muscles are 75% water, the liver is 69% water and living bones are 22% water. A living person is 70% water by weight. On average, a person drinks about five times their own weight of water in a year. In a normal life span, a person consumes about 30,000 L of water. Most of the water in the organic structure in the human body is contained within our cells. The body of a newborn consists of more water (about 75%) than that of an elderly person (50%). A muscular body contains more water than a fat body, as body fat contains less water.[1, 2] Human beings can live without food for several months, but without water for only a few days. Table 1.1 gives the tentative water balance in a human body.

Can we think of a human life without water? Many of us will say no, it's not possible, while a few of us will say yes, but only for a few hours. Both are right. Human beings cannot live long without water. Dehydration (lack of water) kills us faster than starvation (lack of food). A person suffering from dehydration with just a small water loss may display symptoms including irritability, fatigue, nervousness, dizziness, weakness and headaches. The food we eat requires water for cultivation and processing. We use water for entertainment and for sports such as swimming, sailing and rowing. When water becomes ice, we ski or ice skate. Water is required for cooking, washing, bathing, cleaning and recreational activities. Historically, civilizations blossomed around rivers. Equally, those same civilizations were wiped out as sources of fresh water dried up.

Water is the only chemical substance on earth that exists naturally in all three states. It normally exists in its liquid state and does not have smell or taste. It freezes at 0°C and boils at 100°C. Water is an essential life support system for most living beings and the most widely used of all solvents. It is a universal solvent. Its chemical formula is H_2O, meaning that a water molecule contains one oxygen and two hydrogen atoms connected by covalent bonds.

In addition to the standard atoms of hydrogen (H, mass 1) and oxygen (O, mass 16), water contains a very tiny fraction of heavier atoms of hydrogen (deuterium H_2 and tritium H_3) and oxygen (O_{18}), called isotopes. Surface water on earth

TABLE 1.1
Water Balance in a Human Body

Water Intake	Milliliters	Water Out	Milliliters
Liquid drinks	900	Urine	1000
Food	800	Stool	100
Oxidation of food	300	Sweat/lungs	900
Total	2000	Total	2000

follows the water cycle, also called the hydrological cycle. Surface water resources (oceans, rivers, lakes, etc.) undergo evaporation, the water vapors thus formed condense as rain (precipitation), the rain waters run to the rivers/oceans as well as to the interior of the Earth, and the cycle is repeated. The larger fraction of the lighter atom evaporates relatively faster than the tiny heavier atom, while in the precipitation stage the heavier fraction comes down first. This subtle phenomenon results in minute differences in the ratio of the heavy to the light atoms of H and O in water in different parts of the water cycle, as well as in the resultant water and its ultimate storage location. In other words, every drop of water carries its own isotopic fingerprints. The minute differences in the isotopic ratio were traditionally measured using mass spectrometers, and useful information on the origin, pathway and age of the water samples was derived. In recent years laser spectroscopy has begun to be used. Such procedures, called isotopic techniques, form the basis of the science called isotope hydrology.

The specific heat of water is 4.2 J/g°C, which is higher than that of any other common substance. Specific heat is a measure of heat to be added to or released from a substance to change its temperature. This means that 4200 J of heat when added to 1.0 kg of water increases the temperature of water by 1°C. This is a significant amount of heat required to make a small change in the temperature of water. It takes a lot of heat to heat up the water, but once it is hot, it remains hot for a long time. Solids tend to have a lower specific heat capacity. Think of stirring hot soup with a metal spoon. Metal spoons heat up fast. This is because metal has a very low specific heat capacity, so a small amount of heat makes a large temperature change, and water has a high specific heat capacity. Water plays a very important role in temperature regulation, e.g., sweating to dissipate the heat from the human body and control body temperature.

Latent heat of vaporization is another unique property of water useful for several purposes. It is the heat required to convert water into steam at constant temperature. Water molecules form hydrogen bonds with one another. The partial negative charge on the oxygen (O) of one molecule can form a hydrogen bond with the partial positive charge on the hydrogen of other molecules. Water molecules are drawn to other polar ions. Because the bonding electrons are shared unequally by the hydrogen and oxygen atoms, a partial negative charge ($\delta-$) forms at the oxygen end of the water molecule, and a partial positive charge ($\delta+$) forms at the hydrogen ends (Figure 1.1).

Since the hydrogen and oxygen atoms within the molecule carry opposite partial charges, water molecules in the neighborhood are attracted to each other like very

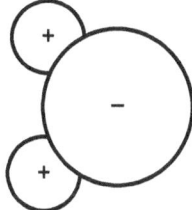

FIGURE 1.1 Charge distribution in H$_2$O molecule.

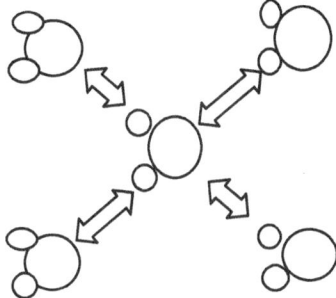

FIGURE 1.2 Hydrogen bonding.

small magnets. The electrostatic attraction between the ð+ hydrogen element and the ð– oxygen element in adjacent molecules is termed hydrogen bonding (Figure 1.2).

Hydrogen bonding causes water molecules to remain together. Hydrogen bonds are relatively weak, but are strong enough to give water several distinctive properties. For example, the *Titanic* sank as a result of hitting an iceberg—a bit of ice floating on the surface of the ocean. The question arises: why does the ice, which is the solid form of water, float on the surface of liquid water? The explanation is: as a result of hydrogen bonding. In the liquid form of water, the hydrogen bonding pulls water molecules together. For this reason, water in liquid form has a relatively compact and dense structure which makes ice float on it.

1.2 WATER ON EARTH

The reality is that 97.5% of water on earth is saline water, leaving only 2.5% as good-quality freshwater. Out of this 2.5%, nearly 70% of the freshwater is frozen in the ice caps of Antarctica and Greenland, and most of the rest is either present as moisture in soil or lies in deep underground reservoirs as groundwater that is not accessible for human consumption. As a result, less than 1% is readily accessible for human consumption. This is often the water found in lakes, rivers, reservoirs and underground sources that are shallow enough to be tapped at an affordable cost. This water is regularly replenished by rain and snowfall and is, therefore, available on a sustainable basis.

Thus, 1% of the world's water supply is a precious commodity, necessary for our survival. And it is unevenly distributed.

There has been a vast growth in the demand for fresh water in recent years, due to rapidly expanding industrial activities as well as the progressive increase in the use of water for irrigation to grow more food for the increasing population. Excessive extraction adversely affects the water table, resulting in groundwater contamination and, in some coastal areas, salt-water intrusion. The depletion of aquifers for irrigation raises questions about the sustainability of water. Added to the overall concern over the sustainable use of water are uncertainties concerning the possible impact of global climate change. The temperature rise due to the greenhouse effect may lead to changing rainfall patterns and evaporation rates.

The global demand for quality water for drinking, sanitation, irrigation and industrial use has been continuously rising, and there has been concern in recent years about water treatment and reuse, to which stringent standards apply. The issues, challenges and opportunities vary from place to place and country to country.

Water resources in several parts of the world are vulnerable due to overutilization and poor management of natural resources as well as ecological degradation and anthropological interventions. Contamination of natural water resources in several areas is increasing day by day due to human abuses, dumping of untreated waste, discharge of untreated industrial effluents, leaching from refuse sites, runoffs from agricultural fields and domestic waste. In view of the scarcity of water resources, it is necessary to understand and develop methodologies for treatment and purification of contaminated water.

The quality of water required by industry varies depending on the type of industry. Among the water-intensive industries are textiles, mining, tanning, automotive, energy, micro-electronics, pharmaceuticals, metal processing, pulp and paper, chemicals and petrochemicals, iron and steel, food and beverages, glass and ceramics, cosmetics and biotechnologies.

Water resources are becoming increasingly scarce worldwide. World water consumption is increasing at more than double the rate of world population growth. Population growth, pollution and climate change, which are all accelerating, are likely to combine to produce a significant decline in water supply in the coming decades. Table 1.2 shows that the largest user of water is agriculture, accounting for 70% of global water withdrawals in 2000, while the respective shares of industrial

TABLE 1.2
Tentative Water Withdrawals in Different Sectors

		Water Consumption as a Share of Total Use (%)		
S. No.	Country Group	Domestic	Agriculture	Industry
1	Low income	6	88	5
2	Middle income	10	70	20
3	High income	15	41	44
4	World	10	70	20

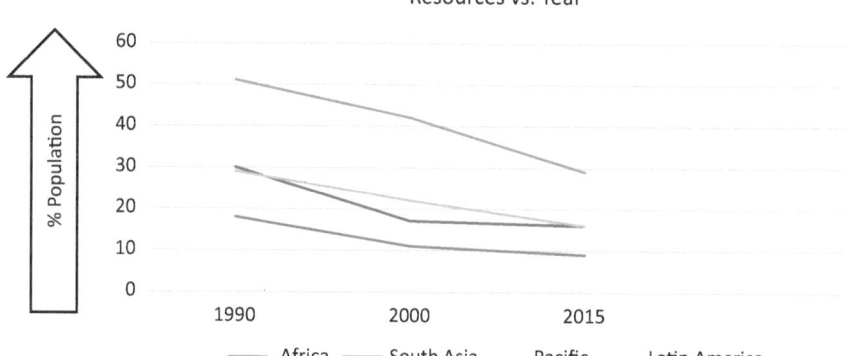

FIGURE 1.3 Percentage population without access to reliable water sources, 1990–2015. *Source*: World Bank, 2003.

and domestic usage were 20% and 10%.[3, 4] Agricultural use dominated globally (up to 70%) and in developing countries, it was more than 85%.

Meanwhile, the share of industrial and domestic use increased with rising country incomes, whereas agricultural use declined. Figure 1.3 shows the percentage of the population without access to reliable water sources, by region, and the predicted values for 2015.[5] Lack of drinking water and sanitation kills about 4500 youngsters a day. Young people are missing school because neither their homes nor their schools have adequate water and sanitation facilities. Therefore, sustainable water is a critical aspect of addressing poverty, inequality and their associated problems. The UN's Millennium Development Goal of ensuring environmental sustainability commits governments to "reduce by half the proportion of people without sustainable access to safe drinking water" by 2015, a goal closely connected with the separate goal of access to sanitation and basic hygiene education.

Countries and regions have different environmental, social and economic conditions and different needs with reference to water use, water quality and the type of technology that may be appropriate in their circumstances. Conventional water treatment technologies seek to remove solid and other contaminants, or to neutralize them, and many treatments have a long history of use in systems for producing water for domestic, industrial and agricultural use.[6] Conventional disinfection methods and standard strategies like chlorination and ozonation may produce harmful by-products on reacting with various constituents of wastewater, which may call for a trade-off between optimum disinfection and harmful by-product formation. The possibility of ready access to safe drinking water is becoming increasingly rare in many places because of overexploitation of existing water resources. The adverse effects of global warming are causing uneven rainfall patterns. Increasing levels of contamination due to rapid industrialization and leakage in water distribution systems are leading to deterioration of water quality. In several places, other factors such as a dearth of funding, poor governance, and lack of availability of trained engineers and skilled labor, are commonly recognized obstacles to establishing regional and national water treatment systems. These problems, together with the lack of success in overcoming obstacles to regional and national water supplies, has led to increased interest in point-of-use water treatment devices at the household and community level. Community- and unit-level point-of-use water treatment units avoid several barriers associated with large-scale installations because they are relatively inexpensive. They are purchased as ready-made units, and constructed using readily available materials, avoiding the need for big capital investment, management systems and governance structures.

The United Nations Committee on Natural Resources notes that about 80 countries comprising 20% of world population are suffering from serious water shortage. In several cases, scarcity of water resources has become the limiting factor for economic and social development. Water shortages are increasingly limiting development options. There are regions that face perennial water shortage. Several regions have problems of higher salinity, and fluoride, nitrate, iron, arsenic and biological contamination in groundwater. The contamination leads to different kinds of issues for their inhabitants.

Water security is important for sustainable development and poverty alleviation. According to Jan Eliasson, Deputy Secretary-General of the United Nations,[7] "The world is experiencing a surge of water-related crises…Rapid urbanization is creating huge pressure on water use and infrastructure, with lasting consequences on human health and urban environments. These changes make water an increasingly scarce and expensive resource, particularly for the poor, the marginalized and the vulnerable. In today's world, we can see how the lack of access to water can fuel conflict and even threaten peace and stability."

A great many people are affected by water-borne diseases every year. So, we need not 'just water', but clean and safe drinking water. The situation may be aggravated by the climate change threat. According to the IPCC Climate Change 2014 Synthesis Report,[8] "Warming of the climate system is unequivocal. The atmosphere and ocean have warmed, the amounts of snow and ice have diminished, sea level has risen…" This rise in sea level and the consequent inundation of low-lying areas, the possible melting of glaciers, the effect on the monsoon, and so on, would also have an impact—direct and indirect—on water security, which is linked to energy security.

1.3   WATER QUALITY

Water is a primary component that supports life. The ability to support life depends on water quality. Adequate supply of water of an appropriate quality is vital for agriculture, domestic use, industries and other requirements. The primary objective of agricultural development has been to maximize yield, at times at the cost of environmental quality. Environmental quality objectives now play a major role in development and management strategies. Human interference has changed ecosystems and may continue to do so. Direct measures of water quality are the concentrations of biological, chemical and physical contaminants. Quality of aquatic habitat involves additional parameters such as temperature, turbidity and dissolved oxygen. Water quality standards and regulations have been established for drinking water, effluent discharge, etc. There are several parameters which play an important role in water quality, such as pH, temperature, dissolved oxygen, turbidity, suspended solids, micronutrients, alkalinity, hardness, salinity, pesticides, organic matter, inorganic contaminants, etc. The following aspects must be considered when addressing water quality issues:

- Type of source of contaminants: natural, anthropogenic or both
- Concentrations: desirable, tolerable, toxic
- Effects on soils, plants or aquatic habitats
- Extent of solubility, volatility, density
- Type of reactions: biological or chemical
- Transport mechanism
- Impact on environment

1.3.1 Trophic States

Water may be classified by trophic state, according to the availability of nutrients or productivity. Trophic classification uses a continuous scale ranging from oligotrophic to mesotrophic to eutrophic. Oligotrophic refers to poor nutrients and having low productivity; mesotrophic means moderate nutrient availability and moderate productivity. The eutrophic is rich in nutrients with high productivity. Oligotrophic water tends to be clear. Eutrophic water tends to be murky. The high organic matter content of eutrophic systems makes them more prone to becoming anoxic. Table 1.3 details the trophic states.

1.3.2 Dissolved Oxygen

The quality of water is strongly related to the amount of oxygen present in it. The oxygen present in natural water decides if it can be a habitat for aquatic species. With a free surface in contact with atmosphere, the exchange between air and water is normally adequate to support a range of species. Under certain conditions, if air–water exchange is not adequate, organisms which can survive with lower oxygen concentration will survive, whereas species that require higher concentrations will not. The amount of dissolved oxygen in the water body varies with depth. In shallow

---

TABLE 1.3

Classification of Water by Trophic State

S. No.	Trophic State	Details
1	Oligotrophic state	Water with less organic matter and minimum biological activity
2	Mesotrophic state	Water with moderate nutrients and more biological productivity
3	Eutrophic state	Water rich in nutrients with high biological productivity
4	Hypereutrophic state	Dark highly productive water
5	Dystrophic state	Low in nutrients, highly colored with dissolved humic organic material

water bodies, vertical mixing carries oxygen to the bottom. The water near the bottom may be anoxic where oxygen demand is higher than the rate of supply.

1.3.3 CONTAMINANTS

Any substance that is present in an amount or concentration that is objectionable or harmful is considered a contaminant. Some contaminants are natural; others are produced and added to nature by human activity. For example, agricultural, municipal and industrial sources have contributed contaminants to both groundwater and surface water. Fertilizers, pesticides, septic tank effluent, animal waste, agricultural sludge and municipal sludge are some of the sources that have potential to add contaminants to water.

1.3.3.1 Biological Contaminants

Microorganisms in drinking water are a potential problem throughout the world. Microorganisms include viruses, bacteria, algae and protozoa. Many of the microorganisms are pathogenic. The common sources of pathogens are human and animal waste.

(a) *Protozoa*

Most parasitic protozoa in humans are less than 50 microns in size. The smallest are 1–10 microns long. Protozoa are unicellular eukaryotes with a nucleus enclosed in a membrane. The common protozoa are *Giardia lamblia*, *Entamoeba histolytica* and *Cryptosporidium*, which cause diarrhea and gastroenteritis.

(b) *Bacteria*

The size range of most of the bacteria is 0.2–2.0 microns. Bacteria may cause cholera (*Vibrio cholera*), typhoid fever (*Salmonella* serogroup Typhi), epidemic dysentery (*Shigella dysenteriae* type 1), etc. Water-borne bacterial infections are normally associated with poor sanitation and hygiene. *Escherichia coli* (*E. coli*) bacteria include many coliform bacteria that inhibit the intestines of warm-blooded animals. Infection is usually through contaminated food or direct contact with the infected individual.

(c) *Viruses*

Viruses are small organisms ranging from 20 to 400 nanometers (nm) in diameter. They can multiply in living cells of animals, plants or bacteria. Enteric viruses cause infection to the gastrointestinal tract of mammals and are excreted in feces. Wherever water supply is contaminated by feces, there is enough potential for transmission of virus. Water-borne viruses of particular concern include hepatitis A, rotaviruses and enteroviruses. Most water-borne viruses infect the intestine and the upper respiratory tract. The most effective method of prevention is good sanitation and hygiene.

1.3.3.2 Chemical Contaminants

Water is known as a universal solvent. A number of inorganic and organic chemicals can be found in water. Solubility and toxicity vary significantly. Effluent discharge from industry and decaying pipe systems are among the known sources of contamination. Some of the common inorganic chemical contaminants, including heavy metals, found in water are arsenic, iron, chromium, lead, mercury, selenium, cadmium, thallium, fluoride, cyanide and nitrate. The term heavy metal is used to refer to metals having atomic weights from that of copper to mercury or having specific gravity more than 4.0. At times, natural mineral deposits may contribute to high concentrations of certain chemicals in water.

1.4 WATER SECURITY

Water, energy and the environment are basic life support systems providing essential inputs for the sustainable development of society. The means of energy production and its consumption have a direct bearing on the environment. Similarly, harnessing water from various resources and its pattern of utilization for agriculture, domestic and industrial purposes has both direct and indirect impact on the environment. It is widely recognized that the future availability of water for domestic, agricultural and industrial requirements is a serious constraint that will adversely affect economic development and human health.

Water security is as important as energy security and food security. A holistic approach to freshwater needs is required, in which desalination and water purification technologies play a very important role. Water is termed safe if it is free from biological contamination (bacteria, etc.) and chemical contaminants (excess salinity, arsenic, fluoride, iron, nitrates, etc.). It is equally important to harness technologies in the areas of bioremediation, breeding drought- and salinity-tolerant crop varieties, and water recycling and reuse. We need advanced cost-effective technologies to recover water from spent streams and effluents for reuse. The revitalization and extension of traditional technologies through integration with new technologies is also necessary. The dissemination and management of knowledge related to the conservation of water in the different agro-ecological regions and its sustainable and equitable use is important.

There is a growing need to develop a science- and technology-based water security system that is economically and environmentally sustainable. Since time immemorial water has been regarded as a free gift from God to mankind, which is one of the important factors responsible for the present worldwide water scarcity and quality crisis. Efforts to address these issues and improve cooperation and understanding on all water-related matters continue worldwide. Technologies for avoiding water wastage and promoting the recycling and reuse of water of appropriate quality have great potential to deal with the water crisis. It is imperative to ensure balanced use of water across different sectors: domestic requirements (drinking and cooking as well as cleaning, washing, etc.), industrial needs, energy production, agriculture and other uses. Sustainable management strategies may involve measures and mechanisms to (i) conserve available water resources; (ii) enhance availability of and access to water resources; (iii) minimize pollution of water resources from industrial and domestic activities; (iv) purify water by mitigation of contaminants; and (v) carry out treatment and recycling of wastewater. Science and technology offer solutions addressing several of the above water-related issues. For example, drip irrigation for agriculture is an efficient water-saving irrigation method. The drippers deliver water directly to the soil adjacent to the root system, which absorbs the water immediately, and thus evaporation loss is minimal.

Since the latter half of the nineteenth century, water technologies have transformed the way people live their lives and where they choose to live. Villages, cities and industries have now developed in many of the arid and water-scarce areas of the world where sea or brackish water (a salt level between fresh and sea water) are available. Effluent is treated with conventional and advanced water technologies. The change is most apparent in parts of the arid Middle East, North Africa, Europe, Asia Pacific, the US and some of the islands of the Caribbean, where the lack of fresh water previously limited development and growth. The requirement to provide fresh water to people in areas with little or no infrastructure was felt during the Second World War. After the war, the need for and potential of desalination and water purification technologies was realized and the technology underwent its first intensive period of development.

1.5   WATER RESOURCE MANAGEMENT

Water resource management, in which water harvesting plays an important role, contributes to water security.

1.5.1   WATER HARVESTING

Uneven distribution of water resources and varied water demand have led to scarcity conditions in several areas. Water scarcity, in turn, has led to overexploitation of groundwater, resulting in rapid depletion of aquifers and lowering of water tables. The overexploitation at times of underground aquifers results in contamination with chemicals such as fluoride and arsenic, salinity and nitrate concentrations beyond the permissible limit.

Emerging and developing economies of the world in particular face severe water scarcity consequences. For example, the estimated utilizable annual water resource in India is 1122 billion m^3, of which 690 billion m^3 and 432 billion m^3 are from surface and groundwater resources respectively.[9] This amount will just be able to meet the expected requirement (1093 billion m^3) by the year 2025. India receives abundant rainfall. The average precipitation is estimated as 4000 billion m^3. However, a sizable portion of the country's precipitation passes on to the sea in the form of surface runoff. Rapid urbanization has further aggravated the problem of runoff, with reduced water permeating to the ground because of the increase in built-up areas and the construction of storm drains. This has reduced the replenishment of groundwater reserves, widening the gap between recharging and drawing from them. Rainwater harvesting measures enable the runoff to be either stored in different forms or made to infiltrate the ground to augment groundwater reserves. Thus, rainwater harvesting is an important tool of water resource management.

Rainwater harvesting involves the collection and storage of rainwater and also other activities aimed at harvesting surface water and groundwater. It implies capturing rainwater during the monsoon, storing it in man-made reservoirs (tanks) or natural reservoirs (aquifers) and using it whenever required. It is also defined as the process of augmenting the natural infiltration of rainwater or surface runoff into the ground by artificial methods or by directly diverting runoff water into existing or disused wells or conserving rainwater by artificially storing and using it for different purposes. The choice and effectiveness of any particular method depends on local factors including the ultimate use of the water. Water can be harvested in situ in tanks, ponds, rooftops, hilltops and other traditional forms of collection. Storage in water aquifers like percolation tanks, check dams, barriers, injection wells, and so on is also possible.

Among several factors that influence the rainwater harvesting potential of a site, the eco-climatic conditions and the catchment characteristics play an important role. The number of annual rainy days also influences the need for and design of rainwater harvesting schemes.

The collection and storage of surface runoff and groundwater has been relegated to relative insignificance with the introduction of large-scale centralized water supply systems using reservoir storage and deep tube wells. In recent years, water harvesting and artificial recharge have picked up. The basic philosophy of rainwater harvesting is "Catch the water where it falls, arrest the water flow and allow recharge".

1.5.2 RECHARGE THROUGH ISOTOPE HYDROLOGY

With the help of environmentally stable (^{2}H and ^{18}O) and radioactive (^{3}H and ^{14}C) isotopes, the source and origin of groundwater recharge can be identified.[10] Isotope techniques in hydrology and water resource management are largely based on the tracer concept. The objective is to directly trace the movement of water molecules in any part of the hydrological cycle and derive information on the transport processes and how such processes are affected by other factors. Isotope techniques have been extensively used to understand the source and mechanism of groundwater pollution. In hilly regions, springs are normally the available source of water for drinking, food production and other uses. These resources are meager and the low rate of discharge during summer causes much hardship to the people. Isotope techniques may be successfully employed to identify the recharge areas of springs, enabling water conservation methods that can rejuvenate dried-up springs (Figure 1.4).

FIGURE 1.4 Recharging of dry springs in mountainous regions of Uttarakhand, India.

At times, it is difficult to ascertain the interconnection between rivers and groundwater. Stable isotopes can be used to estimate the spatial and temporal contributions of rivers to groundwater or vice versa. Surface water systems like lakes and reservoirs are evaporating water bodies, so their contributions to the groundwater can be estimated using isotope techniques.

Identification and estimation of submarine groundwater discharge in coastal regions is important because of the possibility of groundwater loss through the coastline. Natural ^{222}Rn can be used as a tracer to estimate groundwater discharge to coastal areas.

The spatial variation of isotopes in an aquifer can be used to identify areas of modern recharge and flow paths. Tritium is useful for detecting modern recharge. In the absence of modern recharge, carbon-14 measurements can help identify old waters or paleo-waters. They have been used to identify the active recharge area and delineate flow paths in confined aquifers.

1.5.2.1 Groundwater Recharge Investigations

(a) Artificial Recharge of Groundwater from Percolation Tanks

In water-scarce, arid and semi-arid regions, percolation tanks are generally constructed at suitable locations to conserve rainwater as well as to augment groundwater recharge (Figures 1.5A and 1.5B). Environmental isotopes can be used to demarcate the area of influence and quantify the recharge from the constructed percolation tanks to the groundwater.

(b) Identification of Source of Natural Recharge to Groundwater

Stable isotopes (hydrogen-2 and oxygen-18) can be effectively used to identify the source of recharge to the groundwater such as local or distant recharge from precipitation, or recharge from surface water bodies such as rivers and lakes, because the various sources have different isotopic signatures. The stable isotopic composition of groundwater remains unchanged, unless it mixes with other waters of different stable isotopic composition. Generally summer precipitation is enriched with stable isotopes compared to winter precipitation due to temperature effects.

In arid and semi-arid regions, where precipitation is generally low, the challenge is to know whether a given groundwater system is being actively recharged or not. This can be studied using environmental isotopes.

FIGURE 1.5A Percolation tank.

FIGURE 1.5B Well recharged from a percolation tank.

(c) Effect of Storms on Groundwater Recharge

Isotope techniques can be used to quantify storm recharge to groundwater due to the fact that heavy storms are generally depleted in stable isotopic composition compared to normal rains.

(d) Recharge from Lakes and Reservoirs

Surface water systems like lakes and reservoirs are evaporating water bodies and hence their stable isotopic contents are enriched compared to normal groundwater. Subsurface inflow and outflow can be estimated using a water balance and tritium balance model.

(e) Flow Path Delineation and Residence Time Distribution of Groundwater

Using the spatial variation of isotopes in an aquifer, it is possible to identify areas of modern or paleo recharge and flow paths. Radioactive isotope of hydrogen, i.e., tritium, is useful in detecting recent recharges because of its shorter half-life of 12.43 years. The presence of natural tritium in groundwater implies that the recharge is a modern or recent one. If tritium is absent from the groundwater, environmental radioactive carbon-14 (half-life 5730 years) is measured in groundwater, thus identifying old water or paleo-water.

(f) River–Aquifer Interaction

Because of the strong variation in horizontal permeabilities in a river bank, it is often difficult to ascertain the interconnection between river and groundwater using conventional potentiometric contours. Since most rivers originate from higher altitudes, their stable isotopic composition is normally depleted compared to local groundwater. Hence, stable isotopes can be used to estimate the spatial and temporal contribution of a river to groundwater or vice versa.

(g) Interconnection between Aquifers

Knowledge of possible connections between aquifers is useful for evaluating groundwater resources as well as for forecasting the reduction in water quality caused by leakage from the aquifer. In the case of carbonate aquifers, it may be difficult to establish the connection using piezometry and other conventional methods. Stable isotopes of hydrogen (hydrogen-2) and oxygen (oxygen-18) with environmental radioactive isotopes (tritium and carbon-14) can be used to establish connections between aquifers.

(h)   Estimation of Submarine Groundwater Discharge

Identification and estimation of submarine groundwater discharge (SGD) to the coastal regions is important because of the possibility of groundwater loss through the long coastline. Natural radioactive isotope Radon-222 (half-life: 3.8 days), is used as a tracer to estimate SGD.

(i)   Measurement of Groundwater Velocity

Groundwater is a valuable natural resource. The availability of good-quality water is declining all over the world because of increasing dependence on groundwater resources and deterioration in groundwater quality. Groundwater quality is affected by human activities, including industrial discharge, agricultural return flow and urban runoffs. Identification of the source of contaminants and their movement in the subsurface help with the adoption of proper remedial treatment procedures.

Artificial radiotracer techniques are useful tools for measuring the velocity and direction of groundwater flow. Artificially produced isotopes such as bromine-82 (^{82}Br), tritium (^{3}H) and gold-198 (^{198}Au) are normally used as radiotracers for this purpose. The technique involves injecting a suitable radiotracer into a borehole and measuring the dilution rate of the radiotracer in the same borehole over time. The velocity of the groundwater is estimated from the measured concentration data.

1.6   WATER CONTAMINANTS MONITORING THROUGH ISOTOPE TECHNIQUES

Isotope techniques can be used in groundwater pollution studies. Stable isotopes of water (hydrogen-2 and oxygen-18) are used to identify the source, origin and recharge process of groundwater, whereas isotopes such as carbon-13, nitrogen-15, sulfur-34, chlorine-37 and boron-11 can be used to understand the source of chemical pollutants and the existing environmental conditions in the subsurface. Environmental radioactive isotope of hydrogen, i.e., tritium, is used for age dating of modern waters, whereas radioactive isotope of carbon (Carbon-14) helps to determine the age of groundwater on timescales ranging from modern to 40,000 years. Age dating helps to estimate the chronologies of groundwater pollution.

Environmental isotopes can also be used to understand the salinization process in aquifers. At times, the salinity in coastal aquifers may be due to entrapment of old seawater during the Holocene (about 8000 years ago) and late Pleistocene (about 24,000 years ago) periods. Groundwater salinity can also be due to the dissolution of aquifer material as well as entrapped sea water.

Isotope techniques can be applied to identify the source and mechanism of fluoride contamination in groundwater. The source of fluoride in groundwater may be geogenic as well as man-made and it may be due to leaching of fluoride from weathered granite, gneisses and pegmatites and from waste produced by rock-polishing industries located in the vicinity.

These techniques are useful for identifying the source and mechanism of arsenic contamination in groundwater. The findings may indicate that the surface waters, shallow and deep groundwater are free from arsenic contamination, whereas groundwater in intermediate zones may be contaminated with arsenic. The elevated level of arsenic in groundwater may be due to arsenic released from sediment under anoxic conditions in the aquifer.

Thermal power plants dispose of fly ash by backfilling large abandoned coal mines. The chemical present in fly ash percolates and contaminates the groundwater. Isotope techniques can be used to determine the extent of the pollution, and the source and migration rate of pollutants in groundwater downstream of the fly ash disposal site.

Water resources in mountains that originate from snow and glaciers in the upper reaches are now being impacted by climate change. Environmental isotopes can be employed to quantify the various components in the stream flow (snow melt, glacier melt, groundwater and direct runoff).

1.7   MEASUREMENTS OF DISCHARGE RATE IN CANALS AND MOUNTAINOUS RIVERS

Knowledge of the flow rates in canals and mountainous rivers is often useful for water management. Several conventional techniques such as the float velocity method, velocity-area method, discharge meters based on ultrasonic and electromagnetic methods, and dye tracer techniques are commonly used for discharge rate measurements in canals, open channels, streams and rivers. The float velocity method provides a rough estimate of the discharge rate, as it is based on surface velocity rather than mean velocity. Discharge meters based on ultrasonic and electromagnetic methods cannot be used for highly turbulent flows. The fluorescent dye tracer technique is also used to measure discharge rates in canals but has the disadvantage of a significant amount of dye getting adsorbed onto sediments and walls of the canals. The disadvantages and limitations of these conventional techniques can be overcome by the use of radiotracer techniques, which have many advantages including physico-chemical compatibility and high detection sensitivity. Radiotracers facilitate accurate measurement of the discharge rates in canals and mountainous rivers.

1.8   DESALINATION AND WATER PURIFICATION

Desalination is the process by which pure water is produced from saline water using energy. Different forms of conventional energy as well as nuclear and renewable energy sources can be used. Commercially established desalination processes are broadly classified as thermal or membrane processes. The thermal process utilizes heat, which can be waste heat from power plants/process industries, as an energy source for saline water desalination. The membrane process uses nano- and hyper-filtration (also known as reverse osmosis (RO)). Hybrid technologies that combine thermal and membrane processes have the advantage of utilizing the strengths of both processes, such as producing two qualities of desalinated water: (i) distilled water for high-end applications and (ii) potable water for drinking and other uses. Sharing resources and redundancy are possible in hybrid systems. Co-location of desalination and power plants has the advantage of sharing resources and infrastructural facilities.

A water-stressed region needs to ensure water security through integrated water resource management that includes water conservation, seawater desalination, brackish water desalination, wastewater recycling and reuse, water purification and rainwater harvesting. Seawater desalination provides an additional source of fresh water for coastal areas.

Water is a scarce commodity in several parts of the world. Due to the growth of urban populations, better quality of life and the rapid growth of industries, water resources are becoming increasingly polluted. Oil refineries, chemical plants, foundries and pharmaceutical industries, among others, release large volumes of effluents. At times, the effluent may have a high heavy metal content such as cadmium (Cd), chromium (Cr), pesticides, phenolic compounds and other organic compounds. Similarly, there are many rural and remote areas where water sources do exist but are not usable due to the growth of natural vegetation like green algae and macro-foulants such as water hyacinth. Some chemical pollutants such as fluoride and arsenic are released from natural sources, making the water non-potable. Thus, it may be possible that sufficient water is available to cater for the water needs of the human population, but it requires a technology to make it usable. Simple and economically viable advanced water technologies have the potential to go a long way towards meeting water needs in respect of quality and quantity, salvaging this precious resource from the damage caused by man and nature.

REFERENCES

1. Wang, Z.M., Deurenberg, P., Wang, W., Pietrobelli, A., Baumgartner, R.N., and Heymsfield, S.B. 1999. Hydration of fat-free body mass: Review and critique of a classic body composition constant. *American Journal of Clinical Nutrition*, 69: 833–841.
2. Mitchell, H.H., Hamilton, T.S., Steggerda, F.R., and Bean H.W. 1945. The chemical composition of the adult human body and its bearing on the biochemistry of growth. *Journal of Biological Chemistry*, 158: 625–637.
3. Kar, S. and Tewari, P.K. 2013. Nanotechnology for domestic water purification. Chapter 16 of *Nanotechnology in Eco-efficient Construction*, Pacheco-Torgal, F., Diamanti, M.V., Nazari, A., and Granqvist, C. (eds.), Woodhouse Publishing Limited, Cambridge, p. 365.
4. Gleik, P.H. 2000. *The World's Water 2000–2001: The Biennial Report on Fresh Water Resources*. Island Press, Washington, DC.
5. World Bank. 2005. *World development indicators*. Available at http://data.world-bank.org/products/data-books/WDI-2005.
6. OECD. 2011. *Fostering nanotechnology to address global challenges: Water*. Available at http://www.oecd.org/dataoecd/22/58/47601818.pdf. Accessed February 19, 2012.
7. Eliasson, J. 2015. The rising pressure of global water shortages. *Nature*, 6: 517.
8. IPCC Fifth Assessment Report (AR5). 2014. *Climate Change 2014*. Synthesis Report Fortieth Session of the IPCC, IPCC-XL/Doc.21.
9. Bhattacharyya, A., Reddy, J., Ghosh, S.M., and Naika, R. 2015. Water resources in India: Its demand, degradation and management. *International Journal of Scientific and Research Publications*, 5: 346–356.
10. Deodhar, A.S., Ansari, M.A., Sharma, S., Jacob, N., and Singh, G. 2014. Isotope techniques for water resources management. *BARC Newsletter*, 337: 29–35.

2 Membrane Technologies for Water Purification

2.1 INTRODUCTION

Water quality is a major challenge in many parts of the world. In several regions, the majority of wastewater flows untreated or partially treated into water bodies, leading to pollution of water bodies and water sources in many places. Appropriate technology is needed for advanced water treatment and purification.

Treatment technology for water purification depends on the end use of the purified water, and includes:

- Water purification to produce drinking water
- Purification to produce water for reuse for industrial, agricultural or domestic purposes
- Treatment of industrial effluents for safe disposal

A significant amount of work has been carried out on the development of suitable methods of water purification and water reuse. A wide range of conventional water treatment technologies is available for wastewater treatment and for purification of drinking water based on mechanical, thermal, biological, chemical and physical processes. Selection of a certain wastewater treatment methodology depends on several considerations, such as type of wastewater, space availability, the end use of treated water and cost.[1] Mechanical wastewater treatments such as mechanical settlers are used to separate suspended impurities on the basis of the difference in density between the components and water. This technique is at times supported by adding coagulants and flocculants to increase the rate of settling. Thermal processes involve evaporation and distillation techniques. A drawback of thermal processes is the high energy consumption that is required to evaporate the water. Biological treatments include the use of rotating biological contactors, anaerobic filters, sequencing batch reactors and biological aerated filters. The main drawbacks of conventional processes are space requirements, treatment times, extensive testing, monitoring and controlling the parameters, as bacterial growth is influenced by a number of factors.

Physical water treatment processes, particularly membrane-based technologies, have received lots of attention from water technologists.[2, 3] Pressure-driven membrane processes have emerged as an important separation technology increasingly used as an assisted unit operation, with several advantages compared to conventional separation techniques such as distillation, adsorption, absorption, extraction, etc. The membrane works as a semipermeable filter, permeating one component while rejecting another. The separated substances are neither thermally, chemically nor biologically modified. The widespread use of membrane technology can be attributed to several factors such as superior performance, lower cost due to technological advances and water quality regulations that at times cannot be effectively met by conventional treatment processes. Membrane processes are used to improve the quality of drinking water and remove chemical contaminants and microorganisms.

Until 1960, membranes were only used in specialized industrial applications. Membrane applications were almost non-existent due to: (i) reliability issues, (ii) slow transport, (iii) selectivity problems and (iv) cost. These challenges have been addressed by research and development in the field and membrane-based unit operations are now used widely for water purification and other process applications. In the early 1960s, the Loeb–Sourirajan process for preparing anisotropic reverse osmosis (RO) membranes contributed significantly to industrial applications of membrane separation technology.[3] These membranes consist of a thin selective film on a permeable microporous support. The flux of the first Loeb–Sourirajan RO membrane was about ten times higher than that of any existing membrane at that time. This achievement qualified RO as a promising and potentially attractive method for water desalination. The breakthrough resulted in the commercialization of RO and was a major factor in the development of other membrane technologies such as ultrafiltration (UF) and nanofiltration (NF).

High-performance membrane preparation methods, such as interfacial polymerization and multilayer composite casting and coating, were developed based on the Loeb–Sourirajan process. Membranes with selective layers as thin as 0.1 micron (μm) or less were investigated. Methods of packaging large membrane areas into a casing were explored. Different configurations such as tubular, plate-and-frame, spiral-wound, capillary and hollow-fiber modules were developed. By 1980, membrane-based unit operations, such as microfiltration, ultrafiltration, reverse osmosis and electrodialysis, were well established in the industry.

2.2 MEMBRANES

A membrane is a thin barrier between two bulk phases that permits transport of certain ions but retains others depending on their physical and chemical properties. The ability to control the permeation rate of chemical species through the membrane is useful. The main objective in the membrane-based separation process is to allow one component of the feed to permeate through the membrane preferentially, while hindering permeation of others.

A semipermeable membrane separates a feed stream into two streams The part of the feed that has passed through the semipermeable membrane is called the permeate stream, whereas the other part is called the retentate or concentrate stream. The retentate or concentrate contains constituents rejected by the membrane (Figure 2.1). The driving force for the separation process is the pressure difference between the feed and the permeate side, called the transmembrane pressure difference or transmembrane pressure.

The membrane-based separation process has several advantages compared to conventional separation processes: less energy consumption, ecofriendliness, simple

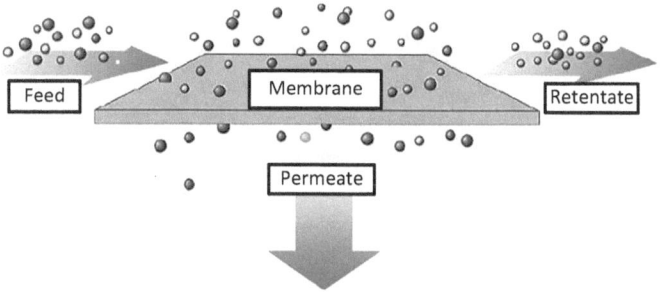

FIGURE 2.1 Semipermeable membrane process.

operation and greater flexibility in system design. However, challenges also need to be addressed: concentration polarization, fouling aspects, flux and selectivity.

As a number of different materials are used to make membranes, they can be broadly classified as either polymeric membranes or inorganic ceramic membranes. Examples of organic polymer membranes that are used in water treatment include: polyethersulfone (PES), polyacrylonitrile (PAN), polyvinylidene fluoride (PVDF), polyphenylsulfone (PPSu), polyetherimide (PEI) and polyamide (PA). Inorganic membranes are prepared from materials such as ceramics, aluminum, high-grade steel, glass and fiber-reinforced carbon. These are used especially if the deployment of polymeric membranes is excluded because of the characteristics of the raw water or if the polymeric membrane surfaces have to be cleaned frequently and intensively due to the wastewater composition. Compared to polymeric membranes, inorganic membranes have high resistance against heat and chemicals, as well as reduced aging and long life. Disadvantages are the higher investment in membrane material and relatively expensive module construction.

Performance aspects of a membrane such as capacity and selectivity are important considerations in determining the economic efficiency of a membrane process. Selectivity is defined as the ability of a membrane to differentiate between the constituents of a mixture and separate one phase from the other. The permeate flow under specific operational conditions describes the capacity of a membrane. Another important feature of a membrane is parameter permeability, which is defined as the quotient from flow and the accompanying transmembrane pressure. The permeability of a membrane is influenced by its condition and the characteristics of the raw water such as temperature, particle size distribution and viscosity.

A membrane differs from a conventional filter because a conventional filter separates particulate suspensions larger than ten (10) microns. A membrane has potential to separate even dissolved solids from the solution. The choice of a particular type of membrane for water purification depends on the specific application desired, such as suspended solids or dissolved solids removal, hardness reduction or ultrapure water production, removal of a particular gas or chemical, etc. The types of commonly used membranes are shown schematically in Figure 2.2. The porous structure of the membranes can be symmetric or asymmetric.

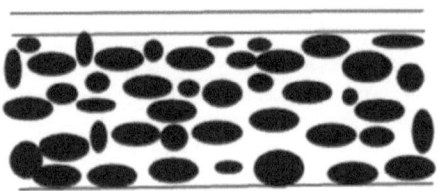

FIGURE 2.2   Thin-film composite (TFC) type of commonly used membranes.

2.2.1 Isotropic (Symmetric) Membranes

Isotropic membranes have uniform structure throughout the entire thickness, with uniform pore size or constant pore size distribution over the whole cross-section of the membrane film. The thickness of isotropic membranes is usually between 30 and 500 μm. The total resistance of the mass transfer depends on the total thickness of the membranes. Hence, a decrease in membrane thickness results in an increased permeation rate.

2.2.1.1 Microporous Membrane

A microporous membrane is like a conventional filter in structure and function. It has a rigid structure with randomly distributed interconnected pores. The pores of microporous membranes are quite small compared to those of conventional filters. The size of the pore varies from 0.01 to 10 μm in diameter. Particles larger than the largest pore size are completely rejected by the membrane. Particles smaller than the largest pores, but larger than the smallest pores, are partially rejected. Particles that are smaller than the smallest pore pass through the membrane. Thus, separation of solutes by microporous membrane is a function of molecular size and pore size distribution. In general, only molecules which differ considerably in size can be separated effectively by microporous membranes, such as in ultrafiltration and microfiltration.

In recent years, interest in membranes made from less conventional materials has increased. Ceramic membranes, a special class of microporous membranes, are being used in ultrafiltration and microfiltration applications where solvent resistance and thermal stability are desirable.

2.2.1.2 Nonporous Membrane

A nonporous membrane consists of a dense film. Component is transported through the dense film by diffusion under the driving force of pressure, concentration or electrical potential gradient. The separation of the components of a mixture depends on their relative transport rate through the membrane. The relative transport rate is determined by their diffusivity and solubility in the membrane material. A nonporous membrane can separate constituents of similar size if their solubility in the membrane material differs. Reverse osmosis uses nonporous membranes for saline water desalination.

2.2.1.3 Electrically Charged Membranes

Electrically charged membranes are microporous or nonporous with fixed positively or negatively charged ions. A membrane with fixed positively charged ions is referred to as an anion-exchange membrane. A membrane containing fixed negatively charged ions is called a cation-exchange membrane. Separation by charged membranes is based on the principle of exclusion of ions of the same charge, and to a much lesser extent on the pore size. The separation is affected by the charge and concentration of the ions in solution. Monovalent ions are excluded less effectively than bivalent ions. Selectivity decreases in solutions of high ionic strength. In electrodialysis, electrically charged membranes are used for processing electrolyte solutions.

2.2.2 ANISOTROPIC (ASYMMETRIC) MEMBRANES

Anisotropic membranes have a gradient in the structure. The cross-section of an asymmetric membrane shows a pore size gradient. There is a relatively dense top layer supported by a more porous sublayer, such as a 0.1–5 μm thick skin layer on a highly porous 100–300 μm thick structure. The skin represents the actual selective barrier of the asymmetric substructure. Its separation properties are determined by the nature of the material or the size of pores in the skin layer. The porous substrate layer serves as a support for the thin top layer, or skin, and has little effect on the separation properties or mass transfer rate of the membrane. The dense surface layer is considered to be responsible for the membrane selectivity. Consequently, the controlled structure of the dense surface layer has become a serious concern in membrane design. Resistance to the mass transfer is mainly determined by the top layer.

Anisotropic membranes are primarily employed in pressure-driven membrane processes such as reverse osmosis, ultrafiltration, gas separation and sometimes in microfiltration. High flux, a reasonable mechanical stability and a very thin selective layer are the unique properties of asymmetric membranes. Two procedures are used to prepare asymmetric membranes. The first method is based on the phase inversion process which leads to an integral structure. In the second method, a thin barrier layer is deposited on a microporous substructure.

The transport rate of species through a membrane is inversely proportional to the membrane thickness. Conventional technology limits manufacture of mechanically strong defect-free films to about 20 μm thickness. A special group of asymmetric membranes are composite membranes. The layers are usually made from different polymers in composite membranes. The separation properties and permeation rates of the membrane are governed by the surface layer. The substructure works as a mechanical support. The advantage of the high flux of anisotropic membranes is so great that many commercial processes use such membranes.

2.3 MEMBRANE TECHNOLOGIES

Microfiltration (MF), ultrafiltration (UF), nanofiltration (NF), reverse osmosis (RO) and electrodialysis (ED) are the established membrane technologies for water treatment and purification. Microfiltration, ultrafiltration, nanofiltration and reverse osmosis are pressure-driven membrane-based filtration processes, whereas electrodialysis is an electrically driven process. Reverse osmosis, nanofiltration, ultrafiltration, microfiltration, and conventional filtration belong to the same family of processes and differ mainly in the pore size of the membrane. The separation mechanism in the case of ultrafiltration and microfiltration is molecular sieving. The pore size in a microfiltration membrane varies from 1000 to 2000 ångström (Å). It requires about 0.1–0.3 bar operating pressure as a driving force and can remove suspended solids and colloidal particles. The pore size in an ultrafiltration membrane varies from 20 to 1000 Å and requires about 0.3 to 5 bar operating pressure. An ultrafiltration membrane can be used to remove bacteria, viruses and dissolved macromolecules such as proteins from solutions. Nanofiltration membranes have pore size in the range of 10–20 Å, require typically 5–5 bar pressure as driving force and can separate micro-sized organic substances and multivalent ions from contaminated water. Discrete pores do not exist in reverse osmosis membranes because the membranes are very dense. Transport occurs through the statistically distributed free volume areas. The pores of the reverse osmosis membrane range from 5 to 10 Å, which is within the range of thermal motion of the polymer chains that form the membrane. The mechanism of transport through the reverse osmosis membrane is governed by the solution-diffusion model, in which solutes permeate the membrane by dissolving in the membrane material and diffusing down a concentration gradient. Separation occurs because of the difference in solubility and mobility of different solutes in the membrane. It typically requires 15–50 bar operating pressure and can remove up to monovalent ions from water.

A membrane has a series of cylindrical capillary pores of diameter (d). The liquid flow rate (q) through a cylindrical capillary pore is given by Poiseuille's Law as:

$$q = \left(\pi d^4 \Delta p \right) / \cdot \left(128 \mu l \right)$$

where Δp is the pressure difference across the pore, μ is the liquid viscosity and l is the pore length.

The membrane flux J, or flow per unit membrane area, is the sum of all the flows through the individual pores. It is given by:

$$J = N \left(\pi d^4 \Delta p \right) / \left(128 \mu l \right)$$

where N is the number of pores per unit area of membrane.

As the pore size for the microfiltration membrane is greater, the flux per unit pressure difference $(J/\Delta P)$ of the microfiltration membrane is higher than that of the ultrafiltration membrane. The flux of the ultrafiltration membrane is higher than that of the nanofiltration and reverse osmosis membranes. This plays an important role in the selection of a membrane process for a particular application.

2.3.1 MICROFILTRATION

Microfiltration is a membrane separation technique in which very fine particles such as suspended solids and colloidal particles are removed from water. The pore size of the microfiltration membrane is typically in the range of 0.1–0.2 μm. Microfiltration (MF) is mainly used for the removal and separation of suspended solids. MF membranes are made from natural or synthetic polymeric materials. Commonly used polymeric materials for MF membranes are cellulose nitrate or acetate, poly-vinylidene difluoride (PVDF), polyamides, polysulfone, polycarbonate, polypropylene, polytetrafluoroethylene (PTFE), etc. Inorganic materials such as metal oxides (alumina), glass, zirconia-coated carbon, etc. are also used for manufacturing MF membranes. Properties such as mechanical strength, temperature resistance, chemical compatibility, hydrophobicity, hydrophilicity, permeability and perm-selectivity play an important role in the choice of membrane materials with respect to particular applications.

2.3.2 ULTRAFILTRATION

Ultrafiltration (UF) is a membrane separation technique in which very fine particles, bacteria, viruses or other suspended matters are separated from raw water. It is also capable of removing suspended solids, proteins and other impurities from effluent or wastewater based on size exclusion mechanisms. Solvents and salts having low molecular weight pass through the pores of the UF membranes while larger molecules are rejected or retained. The primary application of the UF process is the separation of macromolecules.

The UF membrane has pore sizes in the range of 2–100 nanometers (nm), depending upon the type of polymer and procedure used for making them. Since the pores are much smaller than the bacteria, the UF membrane can be used as an absolute barrier to filter out bacteria and suspended solids. By coupling it with an activated charcoal column, other contaminants like color, odor, organics and residual chlorine can also be removed from water. The raw water passes through the membrane under pressure and purified water free from bacteria, suspended solids and high molecular weight organics is produced.

UF membranes are capable of retaining species with molecular weights in the range of 300–500,000 Daltons. UF membranes are described by their nominal molecular weight cut-off (MWCO), which is the smallest molecular weight species for which the membranes have more than 90% rejection. UF is commonly used for removal of macromolecules and microorganisms from raw water. The throughput from UF depends on the physical properties of the membrane, such as permeability and thickness, as well as process variables like feed concentration, system pressure, velocity and temperature.

Characteristics of membrane materials such as porosity, morphology, surface properties, mechanical strength and chemical resistance play an important role in membrane performance. Polymeric materials, such as polysulfone, polypropylene, nylon-6, PTFE, polyvinyl chloride (PVC), acrylic copolymer, etc. are used for making UF membranes. Inorganic materials such as ceramics, carbon-based membranes, zirconia, etc. are also used.

UF separation technology can be designed in two configurations: dead-end filtration and cross-flow filtration. Dead-end filtration is the most basic form of filtration where the feed flow is forced through the membrane and the filtered matter gets accumulated on the surface of the membrane. In dead-end filtration, the accumulated matter on the filter decreases the filtration capacity due to clogging of pores. It requires removal of the accumulated matter. This filtration is a useful technique for concentrating compounds. In cross-flow filtration, a constant flow is maintained along the membrane surface, preventing the accumulation of matter on the surface.

The feed flowing through the membrane is at higher pressure as a driving force for the filtration process and high velocity to create turbulence. This process is referred to as cross-flow, since the feed flow and filtration flow direction are perpendicular to each other. Cross-flow filtration is preferred for liquids having a high concentration of filterable matter.

2.3.3 NANOFILTRATION

Nanofiltration (NF) is a membrane separation technology which can separate up to dissolved multivalent ions from raw water or wastewater. It is a promising choice for applications where high organic removal and moderate inorganic removals are desired. NF gives higher flux than RO. A charged nanofiltration membrane has better selectivity than RO. Since an NF system operates at relatively lower pressure, it has lower energy consumption than a conventional RO system. NF rejects multivalent ions and dissolved materials such as sulfate, phosphate, magnesium and calcium, according to the size and shape of the molecule. The MWCO of a nanofiltration membrane is around 200 Daltons. Some of the specific applications of NF are removal of total organic carbon (TOC), hardness and multivalent ions from surface water, groundwater and wastewater. Membranes used for NF are made of cellulosic acetate or aromatic polyamide having characteristics of giving salt rejection from 95% for bivalent ions to 40% for monovalent salts. NF can typically operate at higher recoveries than RO, thereby conserving total water usage. However, NF is not effective for separation of monovalent ions and low molecular weight organics.

2.3.4 REVERSE OSMOSIS

Reverse osmosis, or RO, is a widely used separation technology based on the membrane process. It removes up to monovalent ions from water. It is capable of removing all metal ions and aqueous salts. Water as solvent passes through the membrane. Dissolved salts as solutes are rejected by the membrane. The operating pressure of RO depends on the osmotic pressure of the solution. Separation of species is a function of the shape and size of the permeating species, their ionic charge, the membrane material properties and its interaction with the permeating species.

Membrane material governs the surface properties of the membrane which in turn has a direct impact on the susceptibility of the membrane to fouling. Depending upon the individual constituents of the water, scaling of the membrane may occur. If the concentration of calcium or magnesium in the water is high, it may cause scaling on the membrane. Hence, it may require appropriate pretreatment of raw water prior to the reverse osmosis membrane.

To understand reverse osmosis, it is necessary to know osmosis. Osmosis is a natural process. If a semipermeable membrane separates two salt solutions of different concentrations, water migrates from the dilute solution through the membrane to the concentrate solution, until the solutions have the same salt concentration. Reverse osmosis involves applying pressure on the concentrated solution side to reverse the natural flow of water, forcing the water to move from the more concentrated solution to the dilute solution. The semipermeable membrane allows water to pass through, but not the salt molecules.

RO technology has the capability to remove dissolved solids, bacteria, viruses and other contaminants contained in the raw water. It operates at ambient temperature and involves no phase change. It is a relatively low energy-consuming process compared to conventional thermal separation processes such as evaporation and distillation. However, it requires higher energy than microfiltration-, ultrafiltration- and nanofiltration-based membrane processes.

RO membranes are made of polymeric material such as cellulose acetate and aromatic polyamide.[4] The membranes are normally of two types: (i) asymmetric or skinned membranes and (ii) thin-film composite (TFC) membranes. The support material is normally polysulfone based whereas the thin film is made from various types of polyamines. The small size of pores does not allow the organic molecules and monovalent/multivalent solutes to pass through the semipermeable membrane along with the water. In practice, more than 99% of inorganic salts and charged organics are rejected by the RO membrane due to the extremely small pores and charge repulsion at the membrane surface.

RO finds extensive applications in production of potable water from seawater, ultrapure water for food processing and electronics industries, pharmaceutical grade water, process water for the chemical, pulp and paper industries, wastewater treatment and water reuse. RO technology has established its potential and gained wider acceptance as a water treatment option for different separation applications. Reasonable cost and the ability to remove more than 99% of inorganic salts as well as organic contaminants with minimal chemical addition make RO an attractive technology for many industrial applications.

2.3.5 ELECTRODIALYSIS

Electrodialysis (ED) is an electro-membrane process. Electro-potential plays an important role in transport of ions through a membrane from one solution to another solution. It is used for different types of separation processes such as separation and concentration of salts, the separation and concentration of monovalent ions from multiple charged components or the separation of ionic compounds from uncharged molecules. ED membranes are normally made of cross-linked sulfonated polystyrene. Anion membranes can be of cross-linked polystyrene containing quaternary ammonia groups. An electrodialysis membrane is usually made in the form of a flat sheet containing about 30–50% water by applying the cation- and anion-selective polymer to a fabric material.

The ED system consists of cation and anion membranes, which are placed in an electric field. The cation-selective membrane permits only the cations and anion-selective membrane permits only the anions. The transport of ions across the membranes results in ion depletion and ion concentration in alternate cells. Electrodialysis is used widely for production of potable water from brackish water, production of ultrapure water, etc.

2.4 MEMBRANE CHARACTERISTICS AND APPLICATIONS

Table 2.1 gives the characteristics of microfiltration, ultrafiltration, nanofiltration and reverse osmosis membranes used in different separation processes.

TABLE 2.1
Membranes Used in Separation Processes

Process	Membrane Type	Typical Pore Size (nm)	Membrane Material	Typical Driving Force (bar)	Applications
Microfiltration	Symmetric microporous	100–200	Cellulose nitrate or acetate, polyvinylidene difluoride (PVDF), polytetrafluoroethylene (PTFE), metal oxides, etc.	0.1–0.3	Removal of suspended solids and turbidity from contaminated water
Ultrafiltration	Asymmetric microporous	2–100	Polysulfone, polypropylene, nylon 6, PTFE, PVC, acrylic copolymer	0.3–5	Removal of organics and microorganisms from water
Nanofiltration	Asymmetric skin type	1–2	Cellulosic acetate, aromatic polyamide	5–15	Separation of bivalent ions and macromolecules from water
Reverse osmosis	Asymmetric skin type	0.5–1.0	Cellulosic acetate, aromatic polyamide	15–50	Separation of monovalent salts and micro-solutes from water

2.5 OTHER MEMBRANE SYSTEMS AND MEMBRANES FOR WATER PURIFICATION

Membrane systems and membranes such as membrane contactor, membrane distillation and biomimetic membranes have good potential in water treatment and purification.

2.5.1 MEMBRANE CONTACTOR

In a membrane contactor, the membrane functions as an interface between two phases but does not control the rate of passage of permeate across the membrane. Membrane contactors are normally shell-and-tube type systems having microporous capillary hollow-fiber membranes. The membrane pores are so small that capillary forces prevent mixing of the phases on either side of the membrane. A liquid gas membrane contactor separates the liquid and gas phases. Membrane contactors can also be used to separate two immiscible liquids (liquid/liquid contactors) or two miscible liquids (usually called membrane distillation). Contactors can be used to selectively absorb one component from a gas mixture into a liquid (gas/liquid contactors).

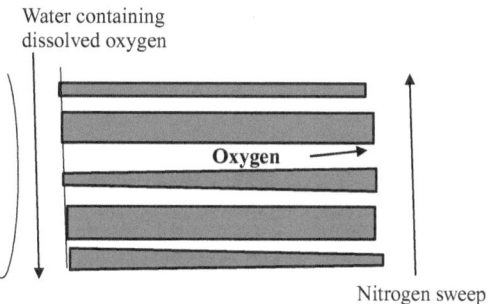

Water containing
dissolved oxygen

Oxygen

Nitrogen sweep

FIGURE 2.3 Typical membrane contactor.

Contactors have a number of advantages, such as high surface area per volume, as compared to a simple liquid gas absorber/stripper or liquid-liquid extractor. The various types of membrane contactors are illustrated in Figure 2.3.

A membrane contactor provides about ten times higher contactor area than an equivalent-sized conventional unit operation. This makes the membrane contactor very compact and lightweight. Its mechanism is based on the physical separation of the counter-flowing phases by the membrane. The membrane area between the two phases is independent of their relative flow rates, so a large flow ratio can be used without producing channeling or flooding or poor phase contact. A small volume of high-value extractants can be used to treat a large volume of low-value feed. Separation of the two phases also eliminates entrainment of one phase into the other, as well as foaming. Unlike traditional contactors, fluids of equal density can be used for the two phases.

The main disadvantage of the contact is related to the nature of the membrane interface. The membrane acts as an additional barrier to transport between the two phases, introducing a resistance to the rate of separation. The membranes can foul during the operation, reducing the permeation rate, or develop leaks, allowing direct mixing of the two phases. The polymeric membranes are quite thin. This introduces a limitation to withstand large pressure differences across the membrane or exposure to harsh solvents and chemicals.

2.5.2 MEMBRANE DISTILLATION

An example of a liquid/liquid membrane contactor is membrane distillation,[5] shown schematically in Figure 2.4. In this process, a warm feed solution flows on one side of the membrane and pure distillate on the other side. Because the solutions are at different temperatures, their vapor pressures are different. As a result, water vapor flows across the hydrophobic membrane. The vapor flux is proportional to the vapor pressure difference between the feed and the permeate. Because of the exponential rise in vapor pressure with temperature, the flux increases sharply as the temperature difference across the membrane is increased.

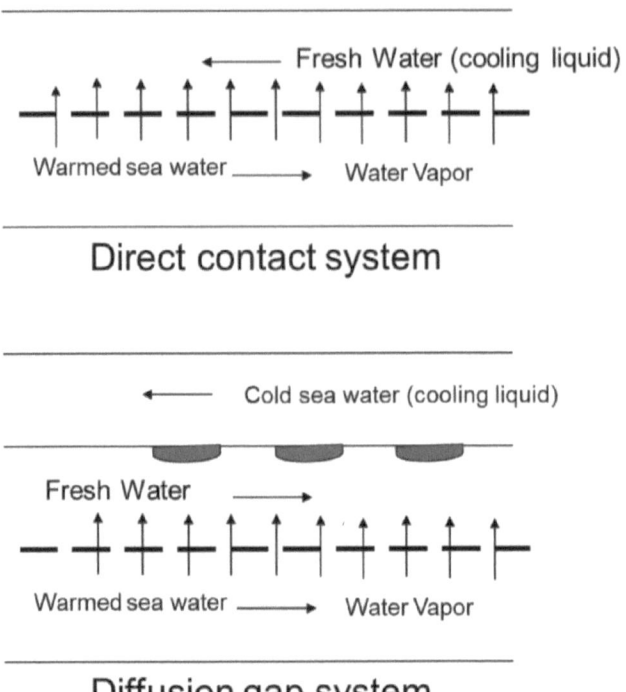

FIGURE 2.4 Membrane distillation process.

Membrane distillation offers several advantages over pressure-driven membrane processes such as reverse osmosis. As membrane distillation is driven by temperature gradient, low-grade waste heat can be used. Unlike RO, membrane distillation systems do not require an energy-consuming high-pressure pump. Membrane distillation can be used for concentrated feed where it is not possible to use the RO system. Increase in osmotic pressure with concentration offers a practical limitation for RO on the concentration of salt in the feed solution to be processed. Research effort is needed focused on technological innovations to reduce costs of membrane modules for membrane distillation systems.

2.5.3 BIOMIMETIC MEMBRANES

The biomimetic membrane concept is based on natural biological systems evolved over billions of years and with proven performance. Biological membranes based on different mechanisms offer efficient separation. Table 2.2 outlines design, development and applications of synthetic membranes.[6] Figure 2.5 suggests that gram-negative bacterial organisms perform size-graded membrane filtration, which is similar to membrane filtration. Structures on the outer membrane surface provide coarse filtration (surface-layer proteins). This is followed by filtration through large

TABLE 2.2

Biological Membrane Separation

Membrane Type	Typical Pore Size	Mechanism	Location in Biological Systems	Relevant Membrane Applications
Surface layer (S-layer) membranes	2–8 nm	Sieving	External membrane surface	Ultrafiltration membrane
Lipid bilayers	Nonporous	Diffusion	Enveloping the cell and cell compartments	Reverse osmosis and forward osmosis membranes
Ionosphere based membranes	Nonporous	Diffusion by carrier	Hydrophobic region of lipid bilayers	Electrode sensors, liquid membranes
Membrane protein facilitated lipid bilayers	0.3–1.5 nm	Diffusion	Transmembrane	Biomimetic desalination membranes, artificial channels
Biological antifouling surface	Not applicable	Surface physicochemical interactions, antifouling chemical signals, surface topography	Membrane surfaces (blood cell membrane, plant and animal skin)	Surface antifouling coating for separations

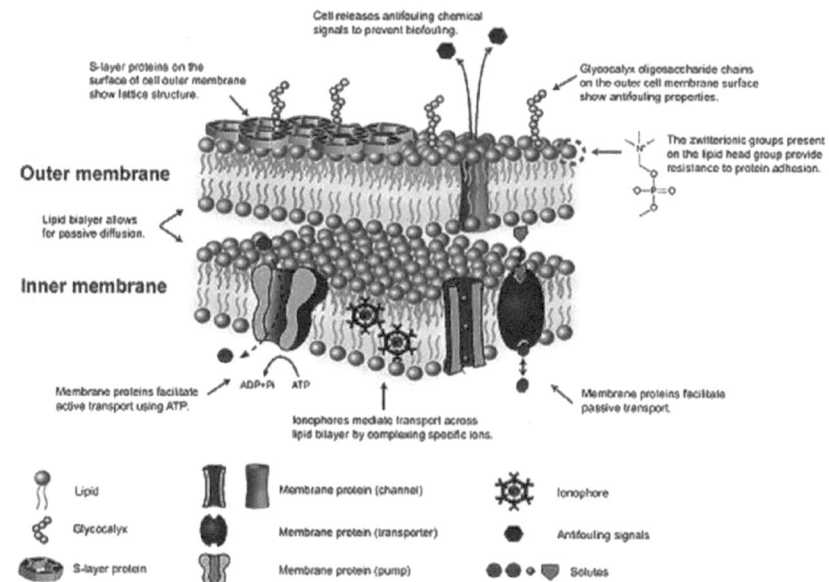

FIGURE 2.5 Separation by biological membrane.

outer membrane channels, which have a particularly high surface area per volume, for macromolecular separations, and inner membrane proteins (or plasma membrane) for specific transport of solutes and water using pumps, channels and transporters. The membrane bilayers permit passive diffusion of water, gases and specific solutes by solution diffusion as well as by carrier mediated diffusion of ionophores in the hydrophobic membrane interior. These cells offer antifouling characteristics to prevent protein deposition on their surface and attachment by other microorganisms.

There are several challenges to overcome with biological membranes: (1) basic understanding of the interaction between functional molecules and matrix materials; (2) engineering approach to the synthesis of biomimetic membranes; and (3) scale-up issues.

2.6 TRANSPORT MECHANISM IN MEMBRANE FILTRATION

The separation in membrane filtration takes place due to differential driving potential across a membrane having selective permeability. Hydrostatic pressure is the differential driving potential for transporting solvent across the ultrafiltration membrane, which is normally used to separate suspended solids, colloids and macromolecules from water. When the solvent flows through the membrane, retained species are concentrated on the membrane surface, creating resistance to the flow. The localized concentration of solute results in precipitation of a solute gel over the membrane. Hence, throughput is a function of the physical properties of the membrane, such as permeability and thickness, and process parameters, such as feed concentration, system pressure, velocity and temperature. Transport mechanisms in membrane filtration can be described as the gel polarization model and the resistance model.

2.6.1 Gel Polarization Model

The gel polarization model assumes that the membranes have a skin that offers minimum resistance to flow, and the asymmetry of the pore virtually eliminates internal pore fouling. The secondary membrane offers the major resistance to flow. The gel layer grows in thickness until the pressure-activated convective transport of solute with solvent towards the membrane surface just equals the concentration gradient of activated diffusive transport away from the membrane surface. Beyond a certain threshold pressure, increase in pressure does not improve the flux since the gel layer grows thicker to offer more resistance to the increased driving force. This is called critical flux.

$$\text{Water flux} \left(J_w \right) = TMP / \left(R_c + R_m \right) \tag{2.1}$$

where TMP is the trans-membrane pressure, R_c is the resistance of the deposited cake and R_m is the hydraulic resistance of the membrane.

Eventually, the concentration at the membrane surface will be high enough to form a gel. In the steady state, the convective transport to the membrane must equal the back diffusive transport away from the membrane.

$$J = -D \left(dC / dx \right) \tag{2.2}$$

where

J: solvent flux through the membrane

C: concentration of solutes or colloids retained in membrane

D: solute diffusivity

x: distance from the membrane surface

Upon integration,

$$J = (D/\delta) \cdot \ln(C_g/C_b) = k \cdot \ln(C_g/C_b) \tag{2.3}$$

where k is the mass transfer coefficient and δ is the boundary layer thickness. C_g and C_b represent maximum solute concentration in the gel layer and concentration of solutes in the bulk of feed respectively. Lower solute concentration (C_b) will have higher threshold pressure since much higher flux is required to transport enough solute to the membrane to begin to form a gel.

2.6.2 Resistance Model

The separation mechanism involves size exclusion as well as adsorption and surface-charge characteristics of membranes. In the absence of a solute, the water flux through a microporous membrane is defined by Darcy's Law, which states that pure solvent flux is directly proportional to the applied pressure differential (ΔP_{app}) and inversely proportional to pure solvent viscosity (μ_w).

$$J_w = \Delta P_{appl} / (R_m \mu_w) \tag{2.4}$$

where R_m is the membrane hydraulic resistance, which is a function of pore size, tortuosity, membrane thickness and porosity.

If the feed solution contains solutes which are retained at the membrane interface, the water flux is generally lower than pure water flux. A number of phenomena have been suggested to account for this flux reduction, such as resistance due to gel layer formation, resistance due to concentration polarization, resistance due to an absorption layer and pore plugging. For a macromolecular solute having high molecular weight at low concentration, the osmotic pressure effect can be neglected. The effect of the gel layer can be represented as

$$J_{uf} = \Delta P_{appl} / (\mu_w (R_m + R_p)) \tag{2.5}$$

where R_p is the resistance due to gel polarization.

The time-dependent case can be represented as

$$J_{uf}(t) = \Delta P_{appl} / (\mu_w (R_m + R_p(t))) \tag{2.6}$$

After testing, if the membrane is thoroughly washed with appropriate washing solution and the pure water flux (J_w) is determined at the same ΔP_{appl}, it may be found to be less than J_w but still greater than J_{uf}. The difference between J_w and J_{uf} is because of the irreversible fouling due to adsorption of solute on the membrane, and this loss in flux can be visualized as additional resistance to the flux (R_a). Hence,

$$J_w = \Delta P_{appl} / \left(\mu_w \left(R_m + R_a \right) \right) \tag{2.7}$$

Incorporating R_a, the equation can be written as:

$$J_{uf}(t) = \Delta P_{appl} / \left(\mu_w \left(R_m + R_a + R_p(t) \right) \right) \tag{2.8}$$

It is noted that J_{uf} reaches an almost constant final flux $J_{uf}(F)$ and the time corresponding to this $J_{uf}(F)$ is $t(F)$. At this stage $R_p(t)$ becomes constant $R_p(F)$.

$$J_{uf}(F) = \Delta P_{appl} / \left(\mu_w \left(R_m + R_a + R_p(F) \right) \right) \tag{2.9}$$

There is also a concentration polarization resulting from the relative rate of solute transport to the membrane surface by convection and the back diffusive solute flux. While both concentration polarization and fouling reduce the membrane flux, they have opposing effects on the observed percent rejection. Another way to distinguish the two phenomena is through their time dependence. Concentration polarization is dependent on parameters such as pressure, temperature, feed concentration and velocity, but is not a function of time. Fouling depends on feed concentration. It is also a function of time.

The change of flux with time due to different kinds of resistances is given in Figure 2.6 for a UF membrane.[7] It shows asymptotic behavior after a particular duration of time.

The mass transfer coefficient, k, can be calculated from the following equation.

$$k = (PR) / (3600 M_B S \cdot \left(\left(1 + m(1-f) M_A \right) / 1000 \right) c (1 - X_3))$$
$$\left(\ln \left((X_2 - X_3) / (X_1 - X_3) \right) \right) \tag{2.10}$$

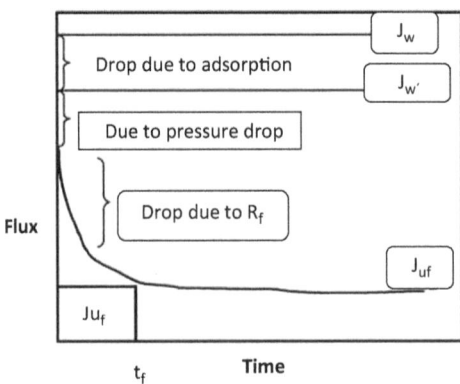

FIGURE 2.6   Change in flux with time for UF membrane.

(PR) refers to the product rate and f refers to the solute separation with reference to the chosen reference solute. M_A and M_B refer to molecular weight of solute and water. S refers to the membrane area, m is the solute molarity and c is the molar density of feed. X_1, X_2 and X_3 refer to mole fraction of solute at bulk, membrane solution interface and membrane permeated product respectively.

2.7 TRANSPORT MECHANISM IN ELECTRODIALYSIS MEMBRANE

Figure 2.7 illustrates the transport mechanism showing concentration and potential gradients in an electrodialysis cell. Chloride ions permeate the anionic membranes containing fixed positive groups, but not the cationic membranes containing fixed negative groups. Sodium ions permeate through the cationic membrane and are stopped by the anionic membrane. This leads to higher salt concentration in alternating compartments while the other compartments have lower salt concentration. The voltage potential drop caused by the electrical resistance takes place entirely across the ion-exchange membrane. The flux of ions across the membranes and hence the productivity of the electrodialysis system can be increased by increasing the current across the stack. In practice, the resistance of the membrane is often small as compared to the resistance of the water-filled compartments, particularly in the dilute compartment where the concentration of ions carrying the current is low. In this compartment, the formation of ion-depleted regions next to the membrane places an additional limit on the current and hence the flux of ions through the membranes. Ion transport through the ion-depleted aqueous boundary layer controls the performance of the electrodialysis system.

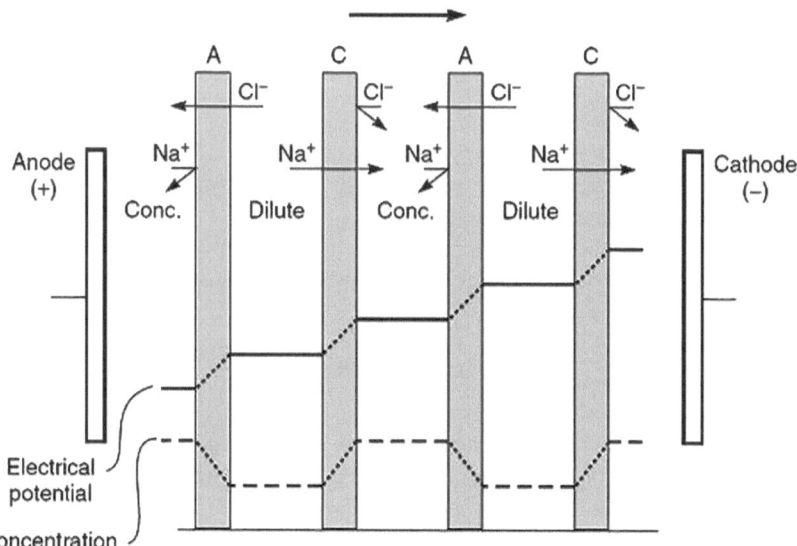

FIGURE 2.7 Concentration and electrical potential gradient profile in electrodialysis (ED) cell.

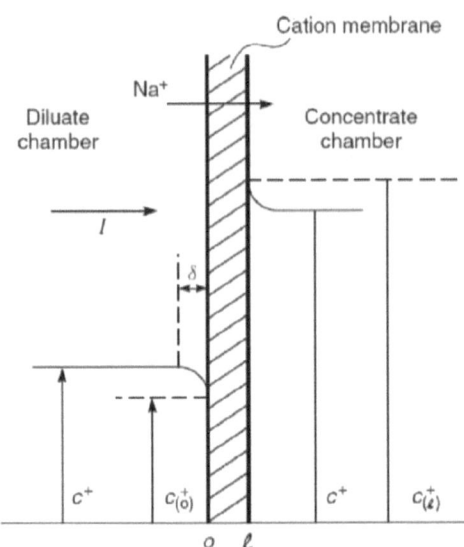

FIGURE 2.8 Concentration gradient of monovalent sodium adjacent to a single cationic membrane in an electrodialysis (ED) stack.

Figure 2.8 shows the concentration gradient of monovalent sodium (Na^+) next to a cationic membrane. An exactly equivalent gradient of anion, such as chloride ion, forms adjacent to the anionic membranes in the stack. The ion gradient formed on the left, dilute side of the membrane can be described by Fick's Law. Thus, the rate of diffusion of cations to the surface is given by:

$$J^+ = D^+ \left(C^+ - C_{(o)}^+ \right) / \delta \tag{2.11}$$

where D^+ is the diffusion coefficient of the cation in water, C^+ is the bulk concentration of the cation in the solution, and $C_{(o)}$ is the concentration of the cation in the solution adjacent to the membrane surface (o).

The rate at which the cations approach the membrane by electrolyte transport is (t^+I/F). The total flux of sodium ions to the membrane surface (J^+) is the sum of these two terms.

$$q = \left(\pi d^4 \Delta p \right) / \cdot \left(128 \mu l \right) \tag{2.12}$$

Transport through the membrane is also the sum of two terms: one due to the voltage difference and the other due to the diffusion caused by the difference in ion concentrations on each side of the membrane. Thus, the ion flux through the membrane can be written as

$$J^+ = \left(t_{(m)}^+ I / F \right) + \left(P^+ \left(C_{(o)}^+ - C_{(l)}^+ \right) / l \right) \tag{2.13}$$

where P^+ is the permeability of the sodium ions in a membrane of thickness (l). The quantity $P^+(C_{(o)}{}^+ - C_{(l)}{}^+)/l$ is quite small as compared to transport due to the voltage gradient, so the above two equations can be combined and simplified to

$$\left(D^+\left(C^+ - C_{(o)}{}^+\right)/\delta\right) + \left(t^+ I / F\right) = \left(t_{(m)}{}^+ I / F\right) \qquad (2.14)$$

For a selective cationic ion-exchange membrane for which $t(m)^+$ is 1, the above equation can be further simplified to:

$$I = \left(FD^+ / \left(1 - t^+\right)\right)\left(C^+ - C_{(o)}{}^+\right)/\delta) \qquad (2.15)$$

This equation has a limiting value when the concentration of the ion at the membrane surface is zero ($C_{(o)}{}^+ = 0$). The limiting value is given by the equation:

$$I_{lim} = \left(FD^+C^+\right)/\left(\delta\left(1 - t^+\right)\right) \qquad (2.16)$$

This limiting current, I_{lim}, is the maximum current that can be employed in an electrodialysis process. If the potential required to produce this current is exceeded, the extra current will be carried by other processes, first by transport of anions through the cationic membrane and, at higher potentials, by hydrogen and hydroxyl ions formed by dissociation of water. Both these undesirable processes consume power without producing any separation, decreasing the efficiency of the process.

The limiting current density for an electrodialysis system is a function of the salt concentration of the feed solution. As the salt concentration in the solution increases, more ions are available to transport current in the boundary layer, so the limiting current density also increases. For this reason, large electrodialysis systems with several electrodialysis stacks in series will operate with different current densities in each stack, reflecting the change in the feed water concentration as salt is removed.

2.8 MEMBRANE STRUCTURES

From the structural point of view, synthetic membranes are classified into five groups: (i) microporous medium; (ii) homogeneous solid film; (iii) asymmetric structure; (iv) electrically charged barrier; and (v) liquid film with selective carriers. However, some membrane structures may combine these features, for example, a membrane can be microporous, asymmetric in structure and carry electrical charges.

2.8.1 NEUTRAL MICROPOROUS MEMBRANES

The neutral microporous films represent a simple form of membrane resembling the conventional fiber filter as far as the mode of separation and the mass transport are concerned. These membranes consist of a solid matrix with pores. Separation of the chemical components is achieved through sieving, with pore diameter and particle size the governing parameters. Microporous membranes can be made from materials like ceramics, graphite, metal or metal oxides, and from different types of polymers. Their structure can be symmetric, implying that the pore diameters do not vary over

TABLE 2.3

Microporous Membranes: Properties and Applications

Membrane Material	Pore Size (μm)	Manufacturing Process	Application
Ceramic, metal or polymer powder	0.1–20	Pressing and sintering of powder	Microfiltration
Homogeneous polymer sheets (Polyethylene (PE), Polytetrafluoroethylene (PTFE))	0.5–10	Stretching of extruded polymer sheets	Microfiltration, artificial blood vessels
Homogeneous polymer sheets (Polycarbonate (PC))	0.02–10	Track etching	Microfiltration
Polymer solution (Cellulose acetate (CA))	0.01–5	Phase inversion	Microfiltration, ultrafiltration

the membrane cross-section. The membranes can also be asymmetrically structured, i.e., the pore diameters may vary from one side of the membrane to the other by a factor of 10–1000. Some of the main properties and applications of microporous membranes are given in Table 2.3.

2.8.1.1 Sintered Membranes

Sintered membranes are made from ceramic materials, glass, graphite and metal powders.[8] Particle size of the powder is an important parameter governing the pore size of the membrane. The membrane can be made in different configurations such as disc, candle or fine-bore tube. Sintered membranes are used for the filtration of colloidal solutions and suspensions. They are also suitable for gas separation and separation of radioactive isotopes.

2.8.1.2 Stretched Membranes

A microporous membrane can be made by stretching a homogeneous polymer film of partial crystallinity. This technique is employed with polyethylene (PE) or polytetrafluoroethylene (PTFE) film extruded from a polymer powder and then stretched perpendicular to the direction of extrusion, leading to relatively uniform pores.[9, 10] These membranes have a very high porosity (up to 90%) and a fairly uniform pore size. Stretched membranes are widely used as microfiltration membranes for acid, caustic solutions and organic solvents. They have replaced the sintered materials used earlier in this application. Stretched membranes can be produced in the form of flat sheets as well as tubes and capillaries. This membrane type has high permeability for gases and vapors due to its high porosity, and because of the hydrophobic nature of the basic polymer, it is impermeable to aqueous solutions. It is also used for removing ethanol[11] from fermentation broths, for pretreatment of seawater in desalination and in medical applications.

2.8.1.3   Capillary Pore Membranes by Track Etching

Microporous membranes with uniform cylindrical pores can be obtained by track etching[12] in two steps. During the first step, a homogeneous thick polymer film is exposed to charged particles in an irradiator. As the particles pass through the film, they leave sensitized tracks where the chemical bonds in the polymer backbone are broken. In the second step, the irradiated film is placed in an etching bath. In this bath, the damaged material along the tracks is preferentially etched forming uniform cylindrical pores.

The density of pores in a track-etched membrane is governed by the residence time in the irradiator. The pore diameter is controlled by the residence time in the etching bath. The minimum pore diameter of these membranes is approximately 0.01 μm. The maximum pore size that can be achieved in track-etched membranes is governed by the etching procedure. Pore size increases with exposure time in the etching medium and the thickness of the film is correspondingly reduced.

Capillary pore membranes are prepared mainly from polycarbonate and polyester films, commercially available in uniform thickness (10–15 μm) which is the maximum penetration depth of collimated particles obtained from irradiator of 0.8–1 MeV energy. Particles with higher energy, up to 10 MeV, can be obtained in an accelerator and are used to irradiate thick polymer films up to 50 μm thickness, or inorganic materials such as mica.[13] Due to their narrow pore size distribution and low tendency to plug, capillary pore membranes made from polycarbonate and polyester have found application on a large scale in analytical chemistry and microbiological laboratories, and in medical diagnostic procedures.[14]

2.8.1.4   Phase Inversion Membranes

Microporous membranes can also be made using the phase inversion technique,[15] in which a polymer is dissolved in a solvent and spread as film of about 20–200 μm thickness. Water is added, causing separation of the homogeneous polymer solution into a solid polymer and liquid solvent phase. The precipitated polymer forms a porous structure containing pores. This type of membrane can be made from the polymer which is soluble in a solvent and can be precipitated in a nonsolvent.[16] By varying the type of polymer, its concentration, the precipitation medium and the precipitation temperature, microporous phase inversion membranes of pore sizes ranging from 0.1 to 20 μm can be made which have desirable chemical, thermal and mechanical properties. It has been reported that phase inversion membranes were originally prepared from cellulosic polymers by precipitation at ambient temperature in an atmosphere of almost 100% relative humidity.[17] Currently, symmetric microporous membranes are also prepared from nylon-66, nomex, polysulfone and polyvinylidene difluoride by precipitation of a cast polymer solution in aqueous liquid.[18] The symmetric, microporous polymer membrane made by phase inversion is widely used in separation processes on an industrial scale.[19] Typical applications range from clarification of turbid solutions, and removal of bacteria or enzymes to detoxification of blood in an artificial kidney.

2.8.2 ASYMMETRIC MICROPOROUS MEMBRANES

An asymmetric membrane consists of a thin selective skin layer (0.1–1 μm) on a porous substructure (100–200 μm thick). The thin skin represents the membrane. Its separation characteristics are governed by the nature of the polymer and the pore size, while the mass transport rate is governed by the membrane thickness. The rate of mass transport is inversely proportional to the thickness of the actual barrier layer. The porous sublayer serves as a support for the thin skin and has little effect on separation characteristics or the mass transfer rate of the membrane. It provides mechanical strength to the membrane. Asymmetric membranes are used in pressure-driven membrane processes such as RO and UF, due to their high mass transfer rate and good mechanical stability.[20] In addition to high filtration rates, asymmetric membranes are fouling resistant. These membranes are surface filters; they retain on the surface all unfiltered materials, which are removed by the shear force applied by the feed solution flowing parallel to the membrane surface.

Two techniques are used to make asymmetric membranes. One uses the phase inversion process; the other is based on achieving a composite structure by depositing a thin polymer film on a microporous substructure.

The development of the asymmetric phase inversion membrane was a major breakthrough in the development and application of the ultrafiltration and reverse osmosis process. The membrane was made from cellulose acetate and yielded fluxes 10–100 times higher than symmetric structures with comparable separation characteristics. Asymmetric phase inversion membranes can be prepared from cellulose acetate and other polymers by the following general preparation technique[21]:

- A polymer is dissolved in an appropriate solvent to form a solution containing 10–30 wt% polymer.
- The solution is spread like a film.
- The film is quenched in a nonsolvent like water or an aqueous solution.

During the quenching process, the homogeneous polymer solution divides into two phases: a polymer-rich solid phase and a solvent-rich liquid phase. The polymer-rich solid phase forms the membrane structure. The solvent-rich liquid phase forms the liquid-filled membrane pores. The pores at the film surface, where precipitation occurs rapidly, are smaller than those in the interior of the film, which leads to the asymmetric membrane structure. There are different variants of this general procedure. Loeb and Sourirajan used evaporation and annealing steps. The evaporation step increased the polymer concentration in the surface of the cast polymer solution. During the annealing step the precipitated polymer film was exposed for a certain time period to hot water of 70–80°C.[22] Descriptions are given of the detailed formation mechanisms of microporous symmetric or asymmetric membranes,[23, 24] with both quantitative and qualitative description and correlation of the various preparation parameters with membrane structures and properties.[25, 26]

2.9   HOMOGENEOUS MEMBRANES

Mass transport in homogeneous membranes occurs by diffusion, so permeability is rather low. Hence, homogeneous membranes should be as thin as possible. The separation of various components in a solution is directly related to their rate of transport within the membrane phase, which is determined by their diffusivity and concentration in the membrane matrix.[27–31]

2.9.1   HOMOGENEOUS POLYMER MEMBRANES

In general, mass transfer is greater in amorphous polymers than in cross-linked polymers.[32] Thus, crystallization and orientation are avoided when high permeability and trans-membrane flux are desired. However, physical properties such as mechanical strength and selectivity of the polymer are adversely affected, and the final product represents a compromise between strength, selectivity and mass transfer rates.[33] Flat sheets and hollow fibers are some of the basic configurations of the membrane.[34, 35] Flat sheets can be prepared by casting from solution, by extruding from a polymer melt, or by blow and press molding. Hollow-fiber configuration is generally made by extrusion with central gas injection.[36] Because of their high selectivity for different chemical components, homogeneous membranes are used in various applications involving the separation of different low molecular weight components with identical or nearly identical molecular dimensions. The important applications of homogeneous polymer membranes are in desalination and gas separation.

2.10   LIQUID MEMBRANE

Liquid membranes use selective "carriers" to transport one of the components selectively at a relatively high rate across the liquid membrane interphase.[37, 38] Two different techniques are used for the preparation of liquid membranes: (1) the selective liquid barrier is stabilized as a thin film by a surfactant in an emulsion-type mixture;[39, 40] or (2) a microporous polymer structure is filled with the liquid membrane phase.[41, 42] In this configuration, the microporous structure provides the mechanical strength and the liquid-filled pores offer the selective separation barrier. Both types are used for the selective removal of heavy metal ions or certain organic solvents from industrial waste streams. Though it is relatively easy to form a thin fluid film, it is difficult to maintain and control this film and its properties during the separation process. In order to avoid breaking the film, reinforcement is necessary to support membrane structure.

2.11   ION-EXCHANGE MEMBRANES

Ion-exchange membranes consist of swollen gels carrying fixed positive or negative charges. The membrane-making procedure and properties of ion-exchange membranes are closely related to ion-exchange resins. There are two types of ion-exchange membranes: (1) cation-exchange membrane having negatively charged groups fixed

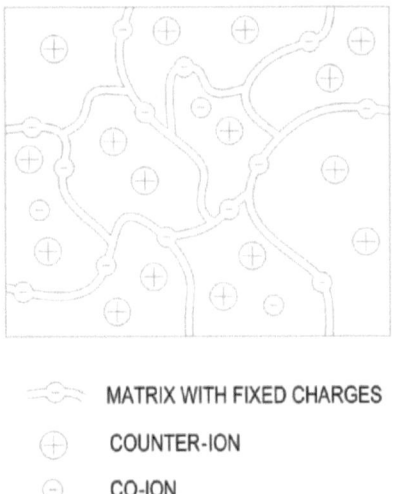

MATRIX WITH FIXED CHARGES

COUNTER-ION

CO-ION

FIGURE 2.9 Cation-exchange membrane having polymer matrix of negative fixed charges, positive counter-ions and negative co-ions.

to the polymer matrix; and (2) anion-exchange membrane having positively charged groups fixed to the polymer matrix. In a cation-exchange membrane, the fixed anions are in electrical equilibrium with mobile cations in the interstices of the polymer. Figure 2.9 shows the matrix of a cation-exchange membrane with fixed anions and mobile cations, which are referred to as counter-ions. In contrast, the mobile anions, called co-ions, are more or less completely excluded from the polymer matrix because of their electrical charge which is identical to that of the fixed ions. Due to the exclusion of the co-ions, a cation-exchange membrane permits transfer of cations only. Anion-exchange membranes carry positive charges fixed on the polymer matrix, thus permeating anions only. The desirable properties of ion-exchange membranes include high perm-selectivity, low electrical resistance, good mechanical stability and high chemical stability.

It is often difficult to optimize the properties of ion-exchange membranes because at times the parameters determining the different properties have conflicting effects. For instance, a high degree of cross-linking improves the mechanical strength of the membrane but also increases its electrical resistance. A high concentration of fixed ionic charges in the membrane matrix leads to low electric resistance but, in general, causes a high degree of swelling combined with poor mechanical stability. The properties of ion-exchange membranes are governed by two parameters: (i) the basic polymer matrix; and (ii) the type and concentration of the fixed ionic moiety. The basic polymer matrix decides the mechanical, chemical and thermal stability of the membrane. Very often the matrix of an ion-exchange membrane consists of hydrophobic polymers such as polystyrene, polyethylene or polysulfone. Although these basic polymers are insoluble in water and show a low degree of swelling, they may become water soluble by the introduction of the ionic moieties. Therefore, the polymer matrix of ion-exchange membranes is often cross-linked. The degree of cross-linking then determines to a large extent the degree of swelling, and the chemical and thermal stability, but also has a major effect on the electrical resistance and the perm-selectivity of the membrane.[43, 44]

The type and the concentration of the fixed ionic charge determine the perm-selectivity and the electrical resistance of the membrane. They also have a significant effect on the mechanical properties of the membrane. The degree of swelling is affected by the fixed charge concentration. The moieties $-SO_3^-$, $-COO^-$, $-PO_3^{2-}$ and $-AsO_3^{2-}$ are used as fixed charges in cation-exchange membranes[45]:

In anion-exchange membranes, fixed charges like $-NH_3^+$ and $=NH_2^+$ are used. The differently charged groups have a significant effect on the ion-exchange behavior of the membrane.

Ion-exchange membranes can be classified into two major categories, heterogeneous and homogeneous. Heterogeneous ion-exchange membranes have some drawbacks, such as relatively high electrical resistance and poor mechanical strength. Homogeneous ion-exchange membranes have significantly better properties in this respect, since the fixed ion charges are distributed homogeneously over the entire matrix. Homogeneous ion-exchange membranes can be made in the following ways[46]:

• Polymerization or polycondensation of monomers.
• Introduction of anionic or cationic moieties into a film by techniques like imbibing styrene into polymer films, polymerizing the imbibed monomer, and then sulfonating the styrene.
• Introduction of anionic or cationic moieties into a polymer chain like polysulfone, followed by dissolving the polymer and casting it into a film.

Heterogeneous ion-exchange membranes consist of fine colloidal ion-exchange particles embedded in an inert binder like polyethylene, phenolic resins or polyvinyl chloride. This can be prepared by impregnating ion-exchange particles into an inert plastic film. Another procedure is dry-molding of inert film-forming polymers and ion-exchange particles and then milling the mold stock. Ion-exchange particles can also be dispersed in a solution containing a film-forming binder, and the solvent evaporated to give an ion-exchange membrane. Similarly, ion-exchange particles can be dispersed in a partially polymerized binder polymer, and the polymerization subsequently completed. Heterogeneous membranes with low electrical resistance contain more than 65% (by weight) cross-linked ion-exchange particles. However, it is difficult to achieve adequate mechanical strength along with low electrical resistance, because the ion-exchange particles swell when immersed in water. Most of the heterogeneous membranes having good mechanical strength generally show poor electrochemical properties.

2.12 COMPOSITE MEMBRANES

In reverse osmosis and pervaporation processes, separation is achieved through a solution-diffusion mechanism in a homogeneous polymer layer. Since the diffusion process in a homogeneous polymer matrix is relatively slow, the membranes

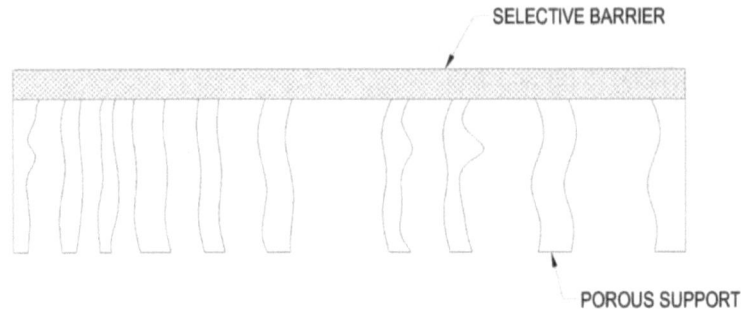

SELECTIVE BARRIER

POROUS SUPPORT

FIGURE 2.10 Asymmetric composite membrane with microporous support structure and selective skin layer.

should be as thin as possible. Therefore, an asymmetric membrane structure is preferable for these processes. Many polymers with the desired selectivity and permeability for the different components in gas mixtures or liquid solutions are not suited to the preparation of asymmetric membranes. This has led to the development of composite membranes. A composite membrane has a 20–100 nm thin dense polymer barrier layer formed over an approximately 100 μm thick microporous film (Figure 2.10). Preparation of the composite membrane consists of two steps: (1) casting of the microporous support, and (2) deposition of the barrier layer on the surface of this microporous support layer. This preparation method for the composite membrane offers several advantages over the integral asymmetric membrane. In an integral asymmetric membrane, the selective barrier layer and the microporous support always consist of the same polymer. In a composite membrane, different polymers are used for the microporous support and the selective barrier layer. This means polymers which show the desired selectivity for a certain separation problem, but have poor mechanical strength or poor film-forming properties, and which are therefore not suited for preparation of integral asymmetric membranes, may well be utilized as the selective barrier in composite membranes. This enhances the choice of available materials for the preparation of semipermeable membranes.

2.12.1 Preparation Method for Composite Membranes

The preparation of composite membrane involves a suitable microporous support, a barrier layer and laminating it to the surface of the support film. The performance of a composite membrane depends on the properties of both the selective barrier layer and the microporous support film.

The preparation of the composite structure involves[47-51]:

* Casting of the barrier layer followed by lamination to the microporous support film.

- Dip coating of microporous support film in a polymer, a reactive monomer or prepolymer solution, to be followed by drying or curing with heat or radiation.
- Gas-phase deposition of the barrier layer of the microporous support film.
- Interfacial polymerization of reactive monomers on the surface of the microporous support film.

Casting an ultrathin film of cellulose acetate in a water surface and transferring the film onto a microporous support was one of the earliest techniques used for preparing composite reverse osmosis membranes for water desalination. Dip coating a microporous support membrane in polymer was also first developed for the preparation of a reverse osmosis membrane. A microporous membrane prepared from mixed cellulose esters was first coated with a protective layer of polyacrylic acid to prevent the barrier layer casting solvent, which consists of cellulose triacetate in chloroform, from dissolving the support membrane. This technique was improved by using a microporous sublayer with better mechanical and thermal stability that was insoluble in the solvent of the barrier layer, such as an open polysulfone ultrafiltration membrane. Dip coating is applied mainly for the preparation of composite membranes to be used in gas separation and pervaporation.[52] Gas-phase deposition of the barrier layer on a dry microporous support membrane by plasma polymerization was also successfully used to make a reverse osmosis membrane.[53]

One of the important techniques for making the composite membrane is the interfacial polymerization of reactive monomers on the surface of a microporous support film. The first membrane to be produced on a large scale with desirable reverse osmosis desalination properties was developed in the early 1970s in the North Star Research Institute under the code name NS100.[54] In the first of two steps, the polyethyleneimine reacts rapidly at the interphase with toluene diisocyanate to form a polyamide surface skin while amine groups below this surface remain unreacted. In the second step, known as heat treatment step, internal cross-linking of the polyethyleneimine takes place. Thus, the final membrane has three distinct layers of increasing porosity: (1) the dense polyamide surface skin which acts as the actual selective barrier, (2) a thin cross-linked polyethylenimine layer which extends into the pores of the support film, and (3) the actual polysulfone support membrane.

2.12.1.1 Sol-Gel Route for Asymmetric Inorganic Membrane

Two sol-gel routes are usually followed[55] based on (i) colloid chemistry in aqueous media, and (ii) chemistry of metal organic precursors in the organic solvent. Both routes are capable of producing porous materials and can be used to make supported ceramic membranes. The porous structure is influenced by the steps involved in the process. The general method for membrane preparation through the sol-gel route is shown in Figure 2.11.[56]

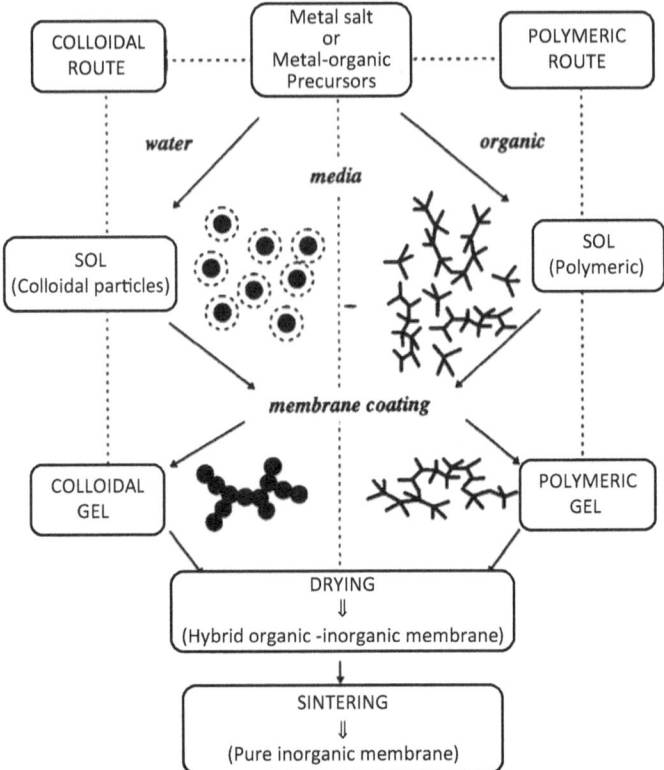

FIGURE 2.11 Sol-gel routes for inorganic membrane preparation.

2.13 MEMBRANE MODULES

Industrial processes often require a large membrane surface area to offer the desired results in the separation process. Therefore, methods of packaging large areas of membrane are important. These packages are called membrane modules. The membranes are cast in the form of flat sheets, tubes, capillary or fine hollow fibers. To accommodate such shapes/structures and to withstand high operating pressure, different types of membrane modules have been developed. The techno-economic factors to be considered in the selection, design and operation of membrane modules include performance, cost of supporting materials and casing (pressure vessel), energy consumption and ease of replacement. Some of the following membrane modules are widely used for industrial applications.

2.13.1 HOLLOW-FIBER MODULES

Hollow-fiber modules are very compact and have a high surface area per unit volume. The diameter of hollow fibers varies from 50 to 3000 μm. Fibers are normally made as a microporous structure having a dense selective layer on the outside or the inside surface. The dense surface layer can be either integral with the fiber or a separate layer coated on the porous support fiber. Several fibers are packed in a tube known as a membrane module. Modules with a surface area of even a few square meters contain kilometers of fibers. Because a module should not contain broken or defective fibers, hollow-fiber production requires stringent quality control and high reproducibility.

Solution spinning and melt spinning are the two methods used to prepare hollow fibers.[57, 58] Solution spinning or wet spinning is a common process, in which a 20–30 wt% polymer solution is extruded and precipitated into a nonsolvent, generally water. Fibers made by solution spinning have the anisotropic structure of membranes. This technique is generally used to make relatively large, porous ultrafiltration and hemodialysis fibers. In the alternative technique of melt spinning, a hot polymer melt is extruded from an appropriate die. It is then cooled and solidified in air before being immersed in a quench tank. Melt-spun fibers are normally denser and have lower fluxes than solution-spun fibers. However, as the fiber can be stretched after it leaves the die, very fine fibers can be made. Melt-spun fibers can be produced at high speed. The technique is normally used to make hollow fine fibers for high-pressure reverse osmosis and gas separation applications. It is also used for polymers such as polytrimethylpentene, which are not soluble in convenient solvents and are difficult to form by wet spinning.

Hollow-fiber membrane modules can be classified into two categories according to the feed arrangement. The first is the shell-side feed design. In this module, a loop or a bundle of fibers is kept in a pressure vessel which is pressurized from the shell side; permeate passes through the hollow fiber. As the fiber wall bears considerable hydrostatic pressure, the fibers usually have small diameters and thick walls, such as 50 μm internal diameter and 100–200 μm outer diameter.

Another type of hollow-fiber module is the bore-side feed type. The fibers in this type of unit are open at both ends, and the feed is circulated through the bore of the fibers. To minimize loss of pressure in the fibers, the diameter of the fiber in this type of module is usually larger than those of the fine fibers used in the shell-side feed system. The hollow fibers are generally made by solution spinning. These modules are popular for ultrafiltration and pervaporation operations. Feed pressures are usually limited to less than 10 bar. Capillary fibers, which are a modified version of hollow fibers, appear promising for several applications where there are concentration polarization and fouling issues.

2.13.2 Plate-and-Frame Module

Plate-and-frame modules[59] are one of the earliest types of membrane configurations, in which membrane, feed spacers and product spacers are placed between two end plates. The feed is sent across the surface of the membrane. A part of it passes through the membrane, enters the permeate channel and makes its way to a central permeate collection manifold. Plate-and-frame modules are popular in electrodialysis and pervaporation systems. This type of configuration is quite popular for effluent treatment applications, particularly for highly fouling feed streams.

2.13.3   Tubular Membrane Module

Tubular modules are generally used for microfiltration and ultrafiltration applications, because of their higher tolerance for membrane fouling. Generally, the membranes are put inside the tubes and tubes have a fiberglass support. A large number of tubes are arranged in series. The permeate stream from each tube is collected in the permeate collection header.

2.13.4   Spiral-Wound Module

In a spiral module, the support fabric, feed spacer and permeate carrier supports the membrane, providing structural integrity, as shown in Figure 2.12. Feed solution moves across the membrane surface. A part of the feed solution passes through the membrane and enters the membrane envelope where it spirals inward to the central perforated collection pipe. The permeate stream and concentrate stream come out of the module. Spiral-wound modules are commonly used by desalination industries for brackish and sea water desalination.

2.13.5   Module Selection

The selection of membrane modules depends on several considerations and design parameters, the most important of which are summarized in Table 2.4.

Concentration polarization control and resistance to fouling are two major factors considered in the selection of a module configuration. Control of concentration polarization is a particularly important issue in membrane separation processes. Another factor is the ease with which membranes can be made and fitted into a module configuration. We can have almost all types of membranes in a plate-and-frame, spiral-wound or tubular module configuration; however, not all types of membrane material can be used to make hollow fine fibers or capillary fibers. The suitability of the module design for high-pressure operation and the relative margin of pressure drops on the feed and permeate sides of the membrane are important parameters in the selection of membrane modules. The types of module used in some of the major commercial separation processes are listed in Table 2.5.

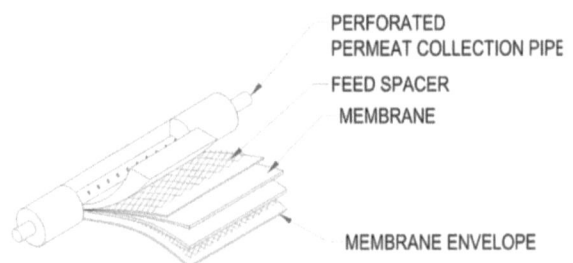

PERFORATED
PERMEAT COLLECTION PIPE

FEED SPACER

MEMBRANE

MEMBRANE ENVELOPE

FIGURE 2.12   Cross-section of spiral-wound module, showing membrane supported by support fabric, feed spacer and permeate carrier.

TABLE 2.4
Membrane Module Configurations Compared

Parameter	Hollow Fiber	Capillary Fiber	Spiral Wound	Plate and Frame	Tubular
Manufacturing cost (US$/m²)	Low	Low	Moderate	High	High
Concentration polarization/ fouling control	Poor	Fair	Moderate	Good	Very good
Permeate-side pressure drop	High	Moderate	Moderate	Low	Low
Limited to specific types of membrane material	Yes	Yes	No	No	No

TABLE 2.5
Commonly Used Module Configurations

Application	Module type
Sea water/brackish water desalination	Spiral-wound modules
Wastewater treatment as well as water recycling and reuse	Tubular, plate-and-frame, capillary and spiral-wound modules

2.14 CONCENTRATION POLARIZATION AND FOULING

Concentration polarization and membrane fouling are the major challenges in membrane separation processes.

2.14.1 CONCENTRATION POLARIZATION

In membrane systems, the feed enters the membrane module, and a permeate enriched with one of the components of the mixture is withdrawn from the other side of the membrane. Because the different components in feed permeate at different rates through the membrane, a concentration gradient is formed on both sides of the membrane. This phenomenon is known as concentration polarization.

The layer of solution adjacent to the membrane surface becomes depleted in the permeating solute on the feed side of the membrane and enriched in this component on the permeate side. Equivalent concentration gradients are also formed for the other components of the feed solution. The concentration polarization reduces the concentration difference across the membrane, decreasing its flux and the membrane selectivity, which can significantly affect membrane performance.

Two approaches are used to describe the effect of concentration polarization. One has its origins in heat transfer analogy. In this approach, resistance to permeation across the membrane and resistance in the fluid layers adjacent to the membrane are treated as resistances in series. The second approach is to model the phenomenon by assuming that a thin layer of unmixed fluid exists between the membrane surface and the well-mixed bulk solution. The concentration gradients, which control concentration polarization, are formed in this layer. This boundary layer film model simplifies the fluid hydrodynamics occurring in membrane modules and contains one adjustable parameter, the boundary layer thickness. This simple model can explain most of the observations.

Membrane processes, such as ultrafiltration and pervaporation, used for the removal of organic solutes from water are both adversely affected by concentration polarization. In ultrafiltration, the low diffusion coefficient of macromolecules produces a concentration of retained solutes several times the bulk solution volume at the membrane surface. At these high concentrations, macromolecules precipitate, forming a gel layer at the membrane surface and reducing the flux. In coupled transport and solvent dehydration by pervaporation, concentration polarization effects are generally modest and controllable.

Cross-, co- and counter-flow schemes are illustrated in Figure 2.13, together with the concentration gradient across the membrane.

The performance of the membrane system can be improved by operating a module in an appropriate flow mode. Normally, counter-flow appears to be promising and beneficial. However, in the case of ultrafiltration and reverse osmosis processes, the selective side of the membrane faces the feed solution, and a microporous support layer faces the permeate solution. Concentration gradient easily builds up in the boundary layer, outweighing the benefit of counter-flow. Counter-flow designs are quite promising in the pervaporation process where the permeate stream is a vapor/gas and permeate-side concentration gradients are more easily controlled because diffusion coefficients of gases are high.

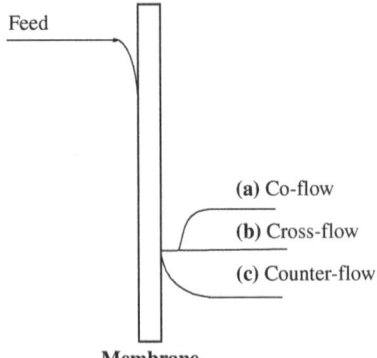

FIGURE 2.13 Concentration gradients of the more permeable component in (a) co-, (b) cross-, and (c) counter-flow cases in a membrane module.

2.14.2 MEMBRANE FOULING

Membrane fouling plays a significant role in deterioration of flux and product quality. The initiation of membrane fouling is due to concentration polarization. The cause and prevention of fouling depend on feed characteristics and appropriate control procedures. Fouling greatly affects the first few modules in the plant, and the last few modules are greatly affected by scaling because they are exposed to the more concentrated feed stream. It is possible to have more than one type of fouling in a membrane system. In general, sources of fouling can be divided into four major categories: scaling, silting, bacteria and organic material. Typical sources of fouling are silt due to organic colloids, corrosion products, algae and fine particulate matter. The silt density index (SDI) is an empirical measurement which gives a fair idea of the probability of fouling due to silt. SDI is the time required to filter a fixed volume of water through a standard 0.45-µm pore size microfiltration membrane. Suspended material in the feed water plugs the microfilter, increasing the sample filtration time and giving a higher SDI. Biological fouling takes place due to deposition of biological matter on the membrane surface. The susceptibility of membranes to biological fouling is a function of membrane composition. Cellulose acetate membranes are an ideal nutrient for bacteria and can be completely destroyed by bacterial attack. Polyamide-based hollow fibers are also somewhat susceptible to bacterial attack. Thin-film composite membranes are quite resistant. Periodic treatment of such membranes with a biocide usually controls biological fouling. Organic fouling is the deposit of organic materials such as oil or grease onto the membrane surface. This type of fouling is common in membrane processes used in industry, particularly effluent streams. Fouling due to scaling and salting are caused by precipitation of dissolved salts in the feed water on the membrane surface. As salts are removed from the feed water, the concentration of ions in the feed increases and may exceed the solubility limit. The salt then precipitates on the membrane surface. The salts that most commonly form scale are calcium carbonate, calcium sulfate, silica complexes, barium sulfate, strontium sulfate and calcium fluoride. Appropriate pretreatment of the feed water helps prevent fouling. Regular cleaning helps remove reversible fouling. Fouling by particulates (silt), bacteria and organics such as oil is generally controlled by suitable pretreatment.

2.15 MATERIALS FOR DIFFERENT MEMBRANE PROCESSES

2.15.1 MATERIALS FOR ULTRAFILTRATION MEMBRANE

A number of materials are used to make ultrafiltration membranes, primarily polyacrylonitrile, poly(vinyl chloride)–polyacrylonitrile copolymers, polysulfone, poly(ether sulfone), poly(vinylidene fluoride) . Aromatic polyamides and cellulose acetate are also used. In general, hydrophilic membranes are more fouling resistant than those made from hydrophobic materials. For this reason, water-soluble polymers such as poly(vinyl pyrrolidone) or poly(vinyl methyl ether) are often added to membrane casting solutions used for hydrophobic polymers such as polysulfone or poly(vinylidene fluoride). During the membrane precipitation step, water-soluble polymer leaches out from the membrane; however, enough remains to make the membrane surface hydrophilic.

The charge on the membrane surface is important. If the membrane surface has a slight negative charge, adhesion of the colloidal gel layer to the membrane is reduced, which helps to maintain a high flux and inhibit membrane fouling. The effect of a slight positive charge on the membrane is the opposite. Charge and hydrophilic character can be the result of the chemical structure of the membrane material or can be applied to the membrane surface by chemical grafting or surface treatment. Choice of treatment depends on the application and the feed characteristics.

2.15.2 MATERIALS FOR REVERSE OSMOSIS MEMBRANE

Reverse osmosis membranes can be grouped into three main categories:

- Seawater desalination membrane operated with 2–5 wt% salt solutions at 30–90 bar pressure.
- Brackish water membrane operated with 0.5–2 wt% salt solutions at 10–30 bar pressure.
- Low-pressure RO membranes used to separate solutes from organic solvent solutions.

Cellulose acetate was one of the initial materials used for reverse osmosis membranes. Higher flux and salt rejection have now been achieved by interfacial composite membranes .[60–63] However, cellulose acetate still maintains a market because of its resistance to degradation by chlorine and other oxidants, which is a major challenge with interfacial composite membranes. A cellulose acetate membrane can tolerate up to 1 ppm chlorine, implying that chlorination can be used for the feed water to deal with biological fouling, is a major advantage with feed streams having significant bacterial loading. The water and salt permeability of cellulose acetate membranes is extremely sensitive to the degree of acetylation of the polymer used to make the membrane.[64–66]

Aromatic polyamide membranes giving high salt rejection and flux have been developed by Toray, Chemstrad (Monsanto) and Permasep (Du Pont) in hollow-fiber configuration.[35, 67, 68] It was observed that high-flux and high-rejection reverse osmosis membranes can be made by interfacial polymerization.[69, 70] The membrane made by Cadotte had salt rejection greater than 99% and flux of 18 gal/ft²day (gfd) at a pressure of 1500 psi for 3.5% sodium chloride solution (synthetic seawater).[71] Current interfacial composite membranes are significantly better. Typical membranes tested with 3.5% sodium chloride solutions have a salt rejection of 99.5% and a water flux of 30 gfd at 800 psi; which is less than half the salt passage of cellulose acetate membranes and twice the water flux.

2.15.3 MATERIALS FOR NANOFILTRATION MEMBRANE

Nanofiltration membranes normally have sodium chloride rejection between 20 and 80% and molecular weight cut-offs of 200–1000 Daltons for dissolved organic solutes. These properties are intermediate between reverse osmosis and ultrafiltration membranes. Reverse osmosis membranes normally have salt rejection of more than 90% and molecular weight cut-off of less than 50; ultrafiltration membranes have salt rejection of less than 5%. Although some nanofiltration membranes are based on cellulose acetate, most of them are based on interfacial composite membranes. The preparation method for these membranes can result in acid groups attached to the polymeric backbone. The rejection of salts by nanofiltration membrane depends on molecular size and Donnan exclusion effects caused by the acid groups attached to the polymer backbone.

The neutral nanofiltration membrane rejects salts in proportion to molecular size, so the extent of rejection is in this order: $Na_2SO_4 > CaCl_2 > NaCl$. The anionic nanofiltration membrane has positive groups attached to the polymer backbone.[72, 73] These positive charges repel positive cations, particularly divalent cations such as Ca^{2+}, while attracting negative anions, particularly divalent anions such as SO_4^{2-}, i.e., salt rejection for $CaCl_2 > NaCl > Na_2SO_4$. The cationic nanofiltration membrane[74, 75] has negative groups attached to the polymer backbone. These negative charges repel negative anions, such as SO_4^{2-}, while attracting positive cations, particularly divalent cations such as Ca^{2+}, implying extent of salt rejection for $Na_2SO_4 > NaCl > CaCl_2$. Low-pressure nanofiltration membranes are normally operated at pressures of 7–15 bar for treatment of 200–5000 ppm salt solutions.

2.15.4 MATERIALS FOR PERVAPORATION MEMBRANES

Most pervaporation membranes are composites formed by solution coating the selective layer onto a microporous support. Microporous polyacrylonitrile coated with a 5–20 μm layer of cross-linked poly(vinyl alcohol) is one of the most commonly used commercial materials.[76] Chitosan[77] and polyelectrolyte membranes such as Nafion[78, 79] have similar properties. Chemically cross-linked poly(vinyl alcohol), formed as a composite membrane by solution casting onto a polyacrylonitrile microporous support, was developed and used for a long time because this membrane has a water/alcohol selectivity of more than 200 and can achieve good separation of water from ethanol or isopropanol solutions.[80] However, the membrane gets swollen and even dissolved by hot acid or base solutions such as hot acetic acid or hot aniline. Membranes stable to such feed solutions can be prepared by plasma polymerization.[81]

2.15.5 MATERIALS FOR ION-EXCHANGE MEMBRANES

Ion-exchange membranes fall into two broad categories, homogeneous and heterogeneous. In homogeneous membranes, the charged groups are uniformly distributed through the membrane matrix. These membranes swell relatively uniformly when exposed to water, the extent of swelling being controlled by their cross-linking density. In heterogeneous membranes, the ion-exchange groups are contained in small domains distributed throughout an inert support matrix, which provides mechanical strength. Heterogeneous membranes can be made by dispersing finely ground ion-exchange particles in a polymer support matrix. Homogeneous membranes can be made by condensation reactions of suitable monomers, such as the phenol–formaldehyde condensation reaction:

The mechanical stability and ion-exchange capacity of these condensation resins are modest. In another approach, a 60:40 mixture of styrene and divinylbenzene is cast onto a fabric web, sandwiched between two plates and heated in an oven to form the membrane matrix. The membrane is then sulfonated with 98% sulfuric acid or a concentrated sulfur trioxide solution. The degree of swelling in the final membrane is controlled by varying the divinylbenzene concentration in the initial mix to control cross-linking density. The degree of sulfonation can also be varied. Anion-exchange membranes can be made from the same cross-linked polystyrene membrane base by post-treatment with monochloromethyl ether and aluminum chloride to introduce chloromethyl groups into the benzene ring, followed by formation of quaternary amines with trimethylamine.

A simple heterogeneous membrane has finely powdered cation- or anion-exchange particles uniformly dispersed in polypropylene. The mechanical properties of these membranes are often poor because of swelling of the relatively large-diameter ion-exchange particles (10–20 μm). A much finer heterogeneous dispersion of ion-exchange particles, and consequently a more stable membrane, can be made with polyvinyl chloride (PVC) plastisol. In this case, a plastisol of approximately equal parts of PVC, styrene monomer and cross-linking agent is prepared in a dioctyl phthalate plasticizing solvent. The mixture is then cast and polymerized as a film. The PVC and polystyrene polymers form an interconnected domain structure. The styrene groups are then sulfonated by treatment with concentrated sulfuric acid or sulfur trioxide to form a very finely dispersed but heterogeneous structure of sulfonated polystyrene in a PVC matrix, providing toughness and strength.

2.16   NEED FOR NANOCOMPOSITE MEMBRANE

The majority of commercially used membranes are made of polymeric material. Interest in membranes made from less conventional but promising material is increasing. Ceramic membranes are a special class of microporous membranes. These membranes are used in ultrafiltration and microfiltration applications where solvent resistance and thermal stability are required. Supported liquid membranes are promising for carrier-facilitated transport processes. To fully utilize the growing opportunities in the field of separation processes, there is strong interest in the identification of new membrane materials which can comply with current requirements and future potential. Durability, mechanical integrity, operating conditions, selectivity, affordability, membrane flux and separation efficiency are important parameters that are considered and evaluated when selecting membrane material for a given separation process. Selectivity, permeation rate, concentration polarization, compaction and anti-biofouling characteristics are important parameters during its operation. For pure polymeric/inorganic materials, there is a general trade-off between permeability and selectivity, with an upper limit. When materials with separation properties near this limit are modified based on the traditional structure–property relation, the resulting polymer has permeability and selectivity tracking along this line instead of exceeding it. On the other hand, inorganic materials possess properties far beyond the upper limit for the organic polymers. Though tremendous improvements have been achieved in tailor-made polymeric structures that enhance separation properties, further progress exceeding the trade-off line seems to present a serious challenge. Similarly, the immediate application of inorganic membranes is hindered by the challenges of making defect-free membranes and the high cost of membrane production. A new out-of-the-box approach is needed to provide an alternate and cost-effective membrane with separation properties well above the currently existing upper limit between permeability and selectivity. The mixed-matrix membrane appears promising and essentially calls for the adoption and usage of composite materials for achieving desirable separation. As regards composite systems, nano-structured materials present an unprecedented opportunity due to the building blocks in this dimension making it possible to design and create unique materials and devices with significantly improved physical, chemical and physico-chemical properties, and flexibility.

Nanocomposites are a solid structure with nanometer-range distances between the different phases that constitute the structure and typically consist of inorganic matrix embedded in the organic phase or organic matrix embedded in the inorganic phase. Nanocomposites demonstrate unique mechanical, chemical, electrical, optical and catalytic properties. Though the idea of causing improvement and enhancing the properties of a material by making multiphase composites is not recent, the application of nanocomposite systems to membrane science and technology is relatively new and still evolving. Using a nanocomposite is aimed at developing an ideal membrane with higher flux, reasonable selectivity and other desirable characteristics according to requirements. Nanocomposite membranes offer preferential permeation for selective transport while acting as a barrier for undesired transport. The quality of the interface between the nanoparticles and the organic polymer is important. It can be altered and improved by chemical modifications of the host polymer matrix and the inorganic nanofillers. A wide variety of nanoparticles have been tested as inorganic nanofillers, including metal oxides (e.g., TiO_2, Al_2O_3, SiO_2, MgO, AgO, Fe_3O_4), pure metals (e.g., nanosilver), zeolites (e.g., ZSM-5, zeolite 4A), nanosized macromer polyoctahedral oligomeric silsesquioxanes (POSS), carbon nanoparticles (e.g., carbon nanotubes, C60 fullerenes) and mineral clays. The nanoscale fillers are normally prepared ex situ and then introduced to the casting mixture. However, it is possible to generate them in situ from precursors. Nanocomposite membranes have great potential in most membrane processes including some novel membrane-based applications.

Nanocomposites can provide the following advantages over conventional membranes:

- Increase in permeability
- Reduced driving force
- Compact system and smaller footprint
- Reduced membrane surface area
- Enhanced mechanical stability
- Enhanced thermal stability

2.17 CASE STUDIES

The following case studies give the salient features of ultrafiltration- and microfiltration-based membrane technologies being used in the field for water purification in rural and remote areas.

2.17.1 ON-LINE WATER PURIFIER BASED ON POLYSULFONE ULTRAFILTRATION MEMBRANE

Bhabha Atomic Research Centre (BARC) in India has developed an on-line water purification device based on an ultrafiltration membrane (Figure 2.14) for removal of microorganisms, suspended solids and high molecular weight organics.[82] The water purification device does not use electricity. Conventional UF membrane units in spiral, capillary and hollow-fiber form have the drawback that they cannot be properly cleaned physically. Their output decreases with time due to deposition of suspended solids. This on-line water purifier is based on an ultrafiltration membrane in cylindrical configuration which is free from these limitations.

The UF membrane device consists of a membrane candle. The membrane candle comprises a porous polypropylene candle laminated with a thin polysulfone-based ultrafiltration membrane. Membrane candles are fitted in a housing unit with an inlet for raw water and an outlet for purified water. The polypropylene candle has about 5 μm pore size. The candle is coated with a polysulfone ultrafiltration membrane in situ using the phase inversion technique and dip-coating method. For this purpose,

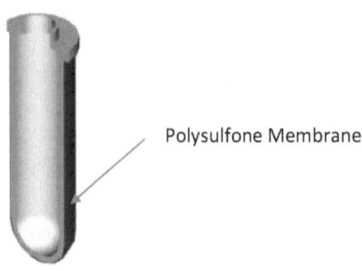

Polysulfone Membrane

FIGURE 2.14   On-line water purification device based on ultrafiltration membrane in candle configuration.

polysulfone polymeric solution with appropriately optimized solvent and additives is used. The total thickness of the membrane coating is about 250 μm and average pore size is less than 10 nm. This pore size is about 50–100 times smaller than the normal sizes of bacteria and hence is capable of acting as an absolute barrier.

The membrane water purification device is designed to work in a dead-end manner, so that no water is wasted. For this purpose, the UF membrane candle is fitted co-axially in the outer housing unit, which is made of suitable plastic or stainless steel. This housing has an inlet for entry of feed water, an outlet for purified product water and provision for easy opening and closing for fixing, cleaning and removal of the membrane candle for routine maintenance.

It can work under normal tap-water pressure. The average candle size is 50 mm diameter and 200 mm length. The filter can supply up to 100 L of purified water daily. It is capable of removing almost all suspended solids and 99.99% of bacteria from the raw water besides reduction of color, odor, organics and residual chlorine. The membrane of the filter is stable in molar acid and alkali as well as residual chlorine. The life of the membrane filter is about 5 years.

This ultrafiltration membrane device for domestic water purification has the following advantages over other devices:

- The device is very effective and almost completely filters out bacteria and turbidity, producing clean water.
- It is compact and portable.
- The device can be easily installed.
- It needs no electricity nor addition of any chemical.
- The device physically filters out bacteria. There will not be any dead, deactivated or decayed bodies of bacteria in the purified water.
- Due to the candle-like cylindrical configuration of ultrafiltration membranes, they can be easily cleaned and reused.
- The operational cost of the device is almost negligible due to no electricity and no chemical requirement. It is almost maintenance free except for occasional cleaning of the membrane surface.
- Polysulfone, the material of preference for making ultrafiltration membranes, has excellent stability with respect to residual chlorine.
- No additional column of activated charcoal is required.
- There is no wastage of feed water.

The technology has been transferred to private parties in different parts of India, several of whom are selling the product in urban as well as remote rural areas.

2.17.2   MEMBRANE-ASSISTED SORPTION-BASED WATER PURIFICATION INSTALLED IN RURAL AREA

The Indian Institute of Technology Jodhpur (IITJ) has been working on solutions to drinking water contamination challenges. A membrane-assisted sorption process which runs on gravity is used for water purification. It does not require electricity, hence is ideal for remote rural locations. It is a low-cost, yet highly effective, decentralized indigenous solution in the form of ultrafiltration (UF) devices for areas that need clean drinking water, particularly for primary-school children in rural parts of Rajasthan State.

These interventions using UF technology have been performed in two districts in north-west India suffering from groundwater fluoride contamination—Sirohi and Jhunjhunu.

2.17.2.1 Modality Followed in Sirohi

Following design and development of the water purification system, installation was carried out, in collaboration with the Sirohi district administration, at the Rajkiya Pradhyamic Vidyalaya on Khadat village, on NH 14, 2 km from Ore in Pindwara region of Sirohi. Figure 2.15 illustrates the initial meeting with the school Principal when the water contamination issue and modality for setting up the water purification unit were discussed. The unit was installed with the help of the district administration, school staff and locally available manpower and is operated by staff and children at the school, ensuring local participation.

Water free from fluoride contamination as well as secondary contaminants like bacteria, viruses, turbidity, suspended solids, color and odor is produced by this UF water purification unit and used by the students (Figure 2.16).

2.17.2.2 Modality Followed in Jhunjhunu

The work was done by IITJ in collaboration with the JJT University, Jhunjhunu. The unit was installed at the government higher secondary school in Dhanuri village, Jhunjhunu, by school staff and local manpower, and is operated by students and staff, ensuring local participation. Figure 2.17 shows the headmaster of the school and members of staff along with IITJ students.

FIGURE 2.15 Primary school at Kharat village in Sirohi district, India, where a membrane-assisted adsorption-based water purification unit is installed.

FIGURE 2.16 Ultrafiltration module (left) and connected elevated water tank (right) at Kharat village in Sirohi district.

FIGURE 2.17   Higher secondary school in Dhanuri village, Jhunjhunu district.

Another UF water purification unit was installed at Rajkiya Uch Madhyamic Vidyalaya (primary higher secondary school) in Luna village, Jhunjhunu. It produces clean drinking water which is used by students and staff (Figure 2.18).

Rural areas have many unique challenges, including availability of electricity, voltage fluctuation, availability of skilled resources and affordability issues. Availability, affordability and accessibility are important parameters for sustainable clean water in rural and remote regions. The technology must be simple, robust and affordable.

FIGURE 2.18 Ultrafiltration module (on the back wall) and JJTU colleagues at school in Luna village, Junjhunu.

2.17.3 Ceramic Membrane-Based Microfiltration

A ceramic membrane-based microfiltration device has been developed by an Indian research laboratory, Central Glass and Ceramics Research Institute (CGCRI). Ceramic membrane microfiltration can be used for removal of suspended solids and turbidity from contaminated water. It can also be used for removal of arsenic contamination from water using the principle of adsorption of arsenic by colloidal media and a membrane-based separation technique using a ceramic microfiltration membrane. The main component of the ceramic membrane-based microfiltration device is a ceramic tube through which water is filtered in cross-flow. The ceramic tube is porous and pore sizes are in the range of 0.1–5 µm (Figure 2.19).[83] Ceramic tubes are housed in a stainless steel chamber. The filtered water is collected as a

FIGURE 2.19 Tubular ceramic elements of ceramic membrane-based microfiltration device.

product. Residual wastewater containing concentrated impurities is collected in separate passages.

In a typical community-size unit, the arsenic- and iron-contaminated water is pumped to a tank containing the adsorbent media and circulated through the tubular ceramic membrane filter module under pressure, producing safe drinking water in the permeate stream.

A typical standard module can supply up to 2500 L/day or 100 L/h purified water. The filtered water has turbidity less than 5 NTU.

REFERENCES

1. Sarkar, B., Chakrabarti, P.P., Vijaykumar, A., and Kale, V. 2006. Wastewater treatment in dairy industries-possibility of reuse. *Desalination* 195: 141–152.
2. Baker, R.W. 2004. *Membrane Technology and Applications*, 2nd Edition, Membrane Technology and Research Inc., Menlo Park, CA.
3. Sirkar, K.K. and Winston Ho, W.S. 1992. *Part I and Part II in Membrane Handbook*, Van Nostrand Reinhold, New York, pp. 1–25.
4. Loeb, S. and Sourirajan, S. 1963. Sea water demineralization by means of an osmotic membrane, in *Saline Water Conversion–II*, Advances in Chemistry Series Number 28, ed. Robert F Gould, American Chemical Society, Washington, DC, pp. 117–132.
5. Lawson, K.W. and Lloyd, D.R. 1997. Membrane distillation. *J. Membr. Sci.* 124: 25.
6. Shen, Y.X., Saboe, P.O., Sines, I.T., Erbakan, M., and Kumar, M. 2014. Biomimetic membranes: A review. *J. Membr. Sci.* 454: 359–381.
7. Ghosh, A.K. 2003. *Ultrafiltration membrane*, PhD thesis, Bombay University, Mumbai, India.
8. Hwang, S.T. and Kammermeyer, K. 1975. *Membranes in Separations*, John Wiley Sons, New York.
9. Druin, M.L., Loft, J.T., and Plovan, S.G. 1974. *Novel open-celled microporous film*. U.S. Patent 3,801,404.
10. Gore, R.W. 1976. *Process for producing porous products*. U.S. Patent 3,962,153.
11. Schneider, K. and van Gassel, T.J. 1984. Membrane distillation. *Chem. Ing. Tech.* 56: 514.
12. Fleischer, R. L., Price, P.B., and Walker, R.M. 1969. Nuclear tracks in solids. *Sci. Am.* 220: 30.
13. Riedel, C. and Spohr, R. 1980. Transmission properties of nuclear track filters. *J. Membr. Sci.* 7: 225.
14. Porter, M.C. 1974. A novel membrane filter for the laboratory. *Am. Lab.* 6: 63.
15. Kesting, R.E. 1971. *Synthetic Polymeric Membranes*. McGraw-Hill, New York.
16. Strathmann, H. 1982. Production of microporous media by phase inversion processes, in *Material Science of Synthetic Membranes*, D.R. Lloyd (ed.), ACS Symposium Series 269, American Chemical Society, Washington, DC, pp. 165–195,
17. Zsigmondy, R. 1992. *Filter and method of producing same*. U.S. Patent 1,421,341.
18. Pall, D. 1982. *Process for preparing hydrophilic polyamide membrane filter media and product*. U.S. Patent 4,340,479.
19. MacLean, D.W. 1958. The MF Millipore filter. *J. New England Water Works Assoc.* 72: 272.
20. Michaels, A.S. 1976. Synthetic polymeric membranes: Practical applications—Past, present and future. *Pure Appl. Chem.* 46: 193–204.
21. Kesting, R.E. 1955. Phase inversion membranes, in *Material Science of Synthetic Membranes*, D.R. Lloyd (ed.), ACS Symposium Series 269, American Chemical Society, Washington, DC, pp. 131–164.

22. Loeb, S. and Sourirajan, S. 1964. *High flow porous membranes for separating water from saline solutions.* U.S. Patent 3,133,132.
23. Merten, U. (ed.). 1966. *Desalination by Reverse Osmosis*, MIT Press, Cambridge, MA.
24. Manjikian, S. 1967. Desalination membranes from organic casting solutions. *Ind. Eng. Chem. Prod. Res. Dev.* 6: 23.
25. Strathmann, H., Scheible, P., and Baker, R.W. 1971. A rationale for the preparation of Loeb-Sourirajan-type cellulose acetate membranes. *J. Appl. Polym. Sci.* 15: 811–828.
26. Strathmann, H., Kock, K., Amar, P., and Baker, R.W. 1975. The formation mechanism of asymmetric membranes. *Desalination* 16: 179–302.
27. Lonsdale, H.K., Merten, U., and Riley, R.L. 1965. Transport properties of cellulose acetate osmotic membranes. *J. Appl. Polym. Sci.* 9: 1341–1362.
28. Stannett, V., Szwarc, M., Ghargava, R.L., Meyer, J.A., Myers, A.W., and Rogers, C.E. 1962. *Permeability of Plastic Films and Coated Paper to Gases and Vapors*, Tappi Monograph Series, No. 23, Technical Association of the Pulp and Paper Industry, New York.
29. Brubaker, D.W. and Kammermeyer, K. 1952. Separation of gases by means of permeable membranes. *Ind. Eng. Chem.* 44: 1465–1474.
30. Stern, S.A. and Walawender, W.P. 1969. Analysis of membrane separation parameters. *J. Sep. Sci.* 4:129.
31. Stannett, V.T., Koros, W.J., Paul, D.R., Lonsdale, H.K., and Baker, R.W. 1979. Recent advances in membrane science and technology. *Adv. Polymer Sci.* 32: 99–151.
32. Crank, J. and Park, G.S. (eds.). 1968. *Diffusion in Polymers*, Academic Press, New York.
33. Stern, S.A. 1976. The separation of gases by selective permeation, in *Membrane Separation Processes*, P. Meares (ed.), Elsevier, Amsterdam, the Netherlands, pp. 295–326.
34. Ward III, W.J., Browall, W.R., and Salemme, R.M. 1976. Ultrathin silicone/polycarbonate membranes for gas separation processes. *J. Membr. Sci.* 1: 99–108.
35. Richter, J.W. and Hoehn, H.H. 1971. *Selective aromatic nitrogen-containing polymeric membranes.* U.S. Patent 3,567,632.
36. McLain, E.A. and Mahon, H.I. 1969. *Permselective hollow fibers and method of making.* U.S. Patent 3,423,491.
37. Ward III, W.J. 1970. Analytical and experimental studies of facilitated transport. *AIChE J.* 16: 405–410.
38. Cussler, E.L. 1971. Membranes which pump. *AIChE J.* 17: 1300–1303.
39. Li, N.N. 1968. *Separating hydrocarbons with liquid membranes.* U.S. Patent 3,410,794.
40. Li, N.N. 1971. Permeation through liquid surfactant membranes. *AIChE J.* 17: 459.
41. Babcock, WC, Baker, R.W., LaChapelle, E.D., and Smith, K.L. 1980. Coupled transport membranes II: The mechanism of uranium transport with a tertiary amine. *J. Membr. Sci.* 7: 71–87.
42. Largman, T. and Sifniades, S. 1978. Recovery of copper (II) from aqueous solutions by means of supported liquid membranes. *Hydrometallurgy* 3: 153–162.
43. Meares, P. 1983. Trends in ion-exchange membrane science and technology, in *Ion-Exchange Membranes*, D.S. Flett (ed.), E. Horwood Ltd., Chichester, England.
44. Spiegler, K.S. and Laird, A.D.K. (eds.). 1980. *Principles of Desalination*, 2nd Edition, Academic Press, New York.
45. Juda, W. and McRae, W.A. 1953. *Ion exchange material and method of making and using.* U.S. Patent 2,636,851.
46. Zschocke, P. and Quellmalz, D. 1985. Novel ion exchange membranes based on an aromatic polyethersulfone. *J. Membr. Sci.* 22: 325–332.
47. Riley, R.L., Lonsdale, H.K., Lyons, C.R., and Merten, U. 1967. Preparation of ultrathin reverse osmosis membranes and the attainment of theoretical salt rejection. *J. Appl. Polym. Sci.* 11: 2143–2158.

48. Riley, R.L., Fox, R.L., Lyons, C.R., Milstead, C.E., Seroy, M.W., and Tagami, M. 1976. Spiral-wound poly(ether/amide) thin film composite membrane systems. *Desalination* 19: 113–126.

49. Cadotte, J.E., King, R.S., Majerle, R.J., and Petersen, R.J. 1981. Interfacial synthesis in the preparation of reverse osmosis membranes. *J. Macromol. Sci. Chem.* A15: 727–755.

50. Yasuda, H. and Marsh, H.C. 1976. Preparation of composite reverse osmosis membranes by plasma polymerization of organic compounds. IV. Influence of plasma–polymer (substrate) interaction. *J. Appl. Polym. Sci.* 20: 543.

51. Cadotte, J.E. 1977. *Reverse osmosis membrane*. U.S. Patent 4,039,440.

52. Henis, J.M.S. and Tripodi, M.K. 1980. A novel approach to gas separations using composite hollow fiber membranes. *Sep. Sci. Tech.* 15: 1059–1068.

53. Yasuda, H. and Lamaze, C.E. 1973. Preparation of reverse osmosis membranes by plasma polymerization of organic compounds. *J. Appl. Polym. Sci.* 17: 201.

54. Rozelle, L.T., Cadotte, J.E., Cobian, K.E., and Kopp, C.V. 1977. Nonpolysaccharide membranes for reverse osmosis: NS-100 membranes, *Reverse Osmosis and Synthetic Membranes*, S. Sourirajan (ed.), National Research Council Canada, Ottawa, Canada, pp. 249–261.

55. Brinker, C.J. and Scherer, G. 1990. *Sol-Gel Science*. Academic Press. New York.

56. Guizard, C. 1996. Sol-gel chemistry and its application to porous membrane processing, in *Fundamentals of Inorganic Membrane Science & Technology*, A.J. Burggraaf and L. Cot (eds.), Elsevier, Amsterdam, the Netherlands.

57. Baum, B., Holley, W. Jr, and White, R.A. 1976. Hollow fibres in reverse osmosis, dialysis, and ultrafiltration, in *Membrane Separation Processes*, P. Meares (ed.), Elsevier, Amsterdam, the Netherlands, pp. 187–228.

58. Moch, I. Jr. 1995. Hollow fiber membranes, in *Encyclopedia of Chemical Technology*, M. Howe-Grant (ed.), 4th Edition, Vol. 13, John Wiley-InterScience, New York, p. 312.

59. Stern, S.A., Sinclaire, T.F., Gareis, P.J., Vahldieck, N.P., and Mohr, P.H. 1965. Helium recovery by permeation. *Ind. Eng. Chem.* 57: 49–60.

60. Riley, R.L. 1991. Reverse osmosis, in *Membrane Separation Systems*, R.W. Baker, E.L. Cussler, W. Eykamp, W.J. Koros, R.L. Riley, and H. Strathmann (eds.), Noyes Data Corp., Park Ridge, NJ, pp. 276–328.

61. Amjad, Z. (ed.). 1993. *Reverse Osmosis*, Van Nostrand Reinhold, New York.

62. Parekh, B. (ed.). 1988. *Reverse Osmosis Technology*, Marcel Dekker, New York.

63. Petersen, R.J. 1993. Composite reverse osmosis and nanofiltration membranes. *J. Membr. Sci.* 83: 81.

64. Reid, C.E. and Breton, E.J. 1959. Water and ion flow across cellulosic membranes. *J. Appl. Polym. Sci.* 1: 133.

65. Rosenbaum, S., Mahon, H.I., and Cotton, O. 1967. Permeation of water and sodium chloride through cellulose acetate. *J. Appl. Polym. Sci.* 11: 2041.

66. Lonsdale, H.K. 1966. Properties of cellulose acetate membranes, in *Desalination by Reverse Osmosis*, M. Merten (ed.), MIT Press, Cambridge, MA, pp. 93–160.

67. Endoh, R., Tanaka, T., Kurihara, M., and Ikeda, K. 1977. New polymeric materials for reverse osmosis membranes. *Desalination* 21: 35.

68. McKinney, R. and Rhodes, J.H. 1971. Aromatic polyamide membranes for reverse osmosis separations. *Macromolecules* 4: 633.

69. Cadotte, J.E. 1985. Evaluation of composite reverse osmosis membrane, in *Materials Science of Synthetic Membranes*, D.R. Lloyd (ed.), ACS Symposium Series Number 269, American Chemical Society, Washington, DC.

70. Larson, R.E., Cadotte, J.E., and Petersen, R.J. 1981. The FT-30 seawater reverse osmosis membrane-element test results. *Desalination* 38: 473.

71. Cadotte, J.E. 1981. *Interfacially synthesized reverse osmosis membrane*. U.S. Patent 4277344.

72. Zhong, S., Widjojo, N., Chung, T.-S., Weber, M., and Maletzko, C. 2012. Positively charged nanofiltration (NF) membranes via UV grafting on sulfonated polyphenylene sulfone (sPPSU) for effective removal of textile dyes from wastewater. *J. Membr. Sci.* 417–418: 52–60.

73. Wang, T., Yang, Y., Zheng, J., Zhang, Q., and Zhang, S. 2013. A novel highly permeable positively charged nanofiltration membrane based on an anoporous hyper-crosslinked polyamide barrier layer. *J. Membr. Sci.* 448: 180–189.

74. Lin, S.W., Sicairos, S.P., and Navarro, R.M.F. 2007. Preparation, characterization and salt rejection of negatively charged polyamide nanofiltration membranes. *J. Mex. Chem. Soc.* 51: 129–135.

75. Childress, A. and Chemelimelech, M. 2000. Relating nanofiltration membrane performance to membrane charge (electrokinetic) characteristics. *Environ. Sci. Technol.* 34: 3710–3716.

76. Bruschke, H.E.A. 1988. State of the art of pervaporation, in *Proceedings of the Third International Conference on Pervaporation Processes in the Chemical Industry*, R. Bakish (ed.), Bakish Materials Corp., Englewood, NJ, pp. 2–11.

77. Watanabe, K. and Kyo, S. 1992. Pervaporation performance of hollow-fiber chitosan-polyacrylonitrile composite membrane in dehydration of ethanol. *J. Chem. Eng. Jpn.* 25: 17–21.

78. Wenzlaff, A., Boddeker, K.W., and Hattenbach, K. 1985. Pervaporation of water–ethanol through ion exchange membranes. *J. Membr. Sci.* 22: 333–344.

79. Cabasso, I. and Liu, Z.-Z. 1985. The permselectivity of ion-exchange membranes for non-electrolyte liquid mixtures: I. Separation of alcohol/water mixtures with nafion hollow fibers. *J. Membr. Sci.* 24: 101–119.

80. Athayde, A.L., Baker, R.W., Daniels, R., Le, M.H., and Ly, J.H. 1997. Pervaporation for wastewater treatment. *CHEMTECH* 27: 34–39.

81. Ellinghorst, G., Steinhauser, H., and Hubner, A. 1992. Improvement of pervaporation plant by choice of PVA or plasma polymerized membranes, in *Proceedings of the Sixth International Conference on Pervaporation Processes in the Chemical Industry*, R. Bakish (ed.), Bakish Materials Corp., Englewood, NJ, pp. 484–493.

82. Tewari, P.K. 2010. *Desalination and Water Purification Technologies: Technical Information Document*, Bhabha Atomic Research Centre, Mumbai, India.

83. Discussion Meet on Desalination and Water Purification Technologies, P.K Tewari, February 2006. Bhabha Atomic Research Centre, Mumbai, India.

3 Nanotechnology for Water Purification

3.1 INTRODUCTION

Nanotechnology can play an important role in water purification, particularly in the removal of contaminants from water. Nanotechnology-based materials have the potential to offer very efficient water treatment technologies. Several water purification devices that incorporate nanoscale materials are commercially available, and others are being developed. These nanotechnology-based products include catalysts and nanoparticles for water treatment. However, nano-based products face a number of challenges. This chapter deals with the practical applications of nanomaterials in water purification and provides an overview of the different types of nanomaterials, such as metal and metal oxide nanoparticles, carbon nanotubes, zeolites, dendrimers, etc., used for water treatment applications. Different routes of synthesis of nanomaterials are discussed, and attention is drawn to the environmental and health implications of the use of nanomaterials. Nanomaterials have also shown their potential in water quality monitoring through sensing and detection. However, this chapter focuses on water purification applications only.

3.2 NANOMATERIALS AND WATER PURIFICATION

Nanotechnology is a generic and evolving term that encompasses the development of a wide range of materials and products. It is engineering at the molecular level and is the collective term for a range of technologies, techniques and processes involving the manipulation of matter at nanoscale from 1 to 100 nanometers. One nanometer (nm) is equal to one-billionth of a meter, or about the width of six (6) carbon atoms or ten (10) water molecules. A human hair is about 80,000 nm wide and a red blood cell is about 7000 nm wide. Atoms are smaller than 1 nm, whereas many molecules including some proteins range between 1 nm and larger.[1] Nanotechnology refers to the ability to engineer materials precisely at the nanometer level. At nanoscale, the properties of materials such as color, magnetism and conductivity change in unexpected ways. The boundaries between established scientific and technical characteristics fade, leading to the strong interdisciplinary character that is associated with nanotechnology. This results in unique characteristics which can generate a vast array of novel products. The field of nanotechnology has opened up new opportunities in water technologies.

The quality and quantity of water available is uneven and differs significantly across the globe. Hence, appropriate location- and region-specific solutions are a necessity. Over time, water-deficient regions become water stressed. Water-stressed regions are becoming water-starved regions. Issues of global warming and climate change have the potential to lead to uneven distribution of rainfall. These factors in turn lead to increasing levels of contamination and at times, detection of new types of contamination. The desirable and permissible limits of contaminants in safe drinking water are evolving with time and becoming ever more stringent. For example, the maximum permissible value for arsenic in drinking water as recommended by the World Health Organization (WHO) has been reduced from 200 to 10 ppb over the course of a number of revisions in the last 50 years. The case is the same with lead (Pb), where the limit has been brought down to 10 ppb from 50 ppb.

Nanotechnology is particularly promising in conditions where reactions occur at ionic/atomic/molecular scale in a very selective manner with high efficiency. Because of the significantly high surface-to-volume ratio due to nano-dimensions, the contaminant uptake capacity becomes many times greater. Compared to technologies such as membrane-based treatments, activated carbon, UV-based filtration, electrodialysis and distillation, nano-based water treatment systems offer the following advantages[2]:

- Higher efficiency of removal even at low concentration of adsorbents
- Functionalization capability of nanomaterials leading to specific uptake
- Low waste generation

3.3 NANOMATERIAL FOR WATER PURIFICATION

Water purification using nanomaterial is governed by either adsorptive or reactive mechanisms. The absorptive remediation process removes contaminants (especially metals) by sequestration, whereas the reactive process leads to degradation of contaminants, sometimes all the way to harmless products such as CO_2 and H_2O in the case of organic contaminants. In-situ technologies involve treatment of contaminants in water, whereas ex situ refers to treatment after removal of contaminants. In-situ degradation of contaminants is often preferred over other approaches because it is potentially more cost-effective.[3] However, in-situ remediation requires delivery of the treatment to the contamination and this is a major obstacle to further development of in-situ remediation technologies. Nanotechnology is especially relevant to this issue because of its potential for injecting nanosized (reactive or absorptive) particles into contaminated water. In this manner, it is possible to create either: (i) in-situ reactive zones with nanoparticles which are relatively immobile; or (ii) reactive nanoparticle plumes which migrate to contaminated zones if the nanoparticles are sufficiently mobile.

Nanoparticles have several main properties making them particularly attractive as sorbents. They have a larger surface area than bulk particles, and nanoparticles can be functionalized with chemical groups to increase their affinity towards target compounds. They have high capacity and act as selective sorbent for metal ions. Nanomaterials can be categorized under four different classes: (i) zeolites; (ii) dendrimers; (iii) metal-containing nanoparticles, like metal oxides; and (iv) carbon nanotubes.

3.3.1 ZEOLITES

Zeolites are microporous aluminosilicate minerals. They are used as commercial adsorbents and catalysts. Zeolites are crystalline solids with well-defined structures belonging to the family of microporous solids known as molecular sieves, which have the ability to selectively sieve molecules based on the principle of size exclusion. This is due to the pore structure of molecular dimensions. The maximum size of molecular or ionic species that can enter the pores of a zeolite is controlled by the size of the channels. Zeolites have silicon, aluminum and oxygen in their framework. They have cations, water and other molecules in their pores. Zeolites occur in nature as minerals and are extensively mined in many parts of the world. They are also made synthetically for specific uses. Synthetic zeolites are made from silicon-aluminum solutions or coal fly ash and used as sorbents in column filters. Zeolites are widely used as ion-exchange beds for domestic and commercial water purification, and water softening. In water softening, alkali metals such as sodium or potassium are exchanged by the calcium and magnesium ions from the feed water. Industrial wastewater containing heavy metals can also be treated using zeolites. Zeolites are commonly used for the removal of metal contaminants. Natural zeolites from Mexico and Hungary have the potential to reduce arsenic from drinking water to WHO-standard safe levels.[4] Zeolites made from coal fly ash can also adsorb a variety of heavy metals such as lead, copper, zinc, cadmium, nickel and silver from wastewater. Under certain conditions, fly ash zeolites can also adsorb chromium, arsenic and mercury. NaP1 zeolites ($Na_6Al_6 Si_{10}O_{32} \cdot 12H_2O$) have a high density of Na (I) ion-exchange sites. They can be synthesized by hydrothermal activation of fly ash with low Si/Al ratio at 150°C in 1.0–2.0 M NaOH solutions. NaP1 zeolites as ion-exchange media have been evaluated for the removal of heavy metals from acid mine wastewater. Successful use of synthetic NaP1 zeolites has been studied for the removal of Cr (III), Ni (II), Zn (II), Cu (II) and Cd (II) from metal electroplating wastewater.[5]

The adsorptive capacity of zeolites is influenced by several factors including their composition, pH value of water, types of contaminants and concentration. For example, lead and copper are more easily adsorbed by fly ash. So, the high concentrations of these metals reduce the amount of cadmium and nickel removal.[6] Zeolite-silver compounds are effective against microorganisms, including bacteria and mold. Additionally, the silver in this compound provides residual protection against regrowth of these biological contaminants. Only zeolites, as such, do not adequately remove organic contaminants. Also, air moisture contributes to saturation of zeolites, making them less effective. Zeolites were reported to have contributed in the development of a high-flux nanocomposite membrane without compromising selectivity.[7]

3.3.2 DENDRIMERS

Dendrons are dendritic wedges that comprise one type of functionality (such as chemical bonding) at their core and another at the periphery. To obtain a dendrimer structure, several dendrons are reacted with a multifunctional core. Using different synthetic strategies, over one hundred compositionally different dendrimer families have been synthesized and over a thousand differentiated chemical surface modifications have been reported.[8, 9] Dendritic polymers exhibit features that make them particularly attractive as functional materials for water purification. These soft nanoparticles, having a size range of 1–20 nm, can be used as recyclable water-soluble ligands for toxic metal ions, radionuclide and inorganic anions. The environmental applications of dendrimers led to effective removal of copper from water by different generations of poly(amidoamine) (PAMAM) dendrimers. The use of dendrimers improved the potential of ultrafiltration for recovering Cu (II) from aqueous solution.[10] The dendrimer-Cu(II) complexes can be effectively separated from aqueous solutions by ultrafiltration. Dendritic polymers can also be used as (i) recyclable unimolecular micelles for recovering organic solutes from water and (ii) scaffolds and templates for the preparation of redox and catalytically active nanoparticles.[11] A water-soluble benzoylthiourea modified ethylenediamine core-polyamidoamine dendrimer was developed for the selective removal and enrichment of toxicologically relevant heavy metal ions. Studies were carried out on the complexation of Co (II), Cu (II), Hg (II), Ni (II), Pb (II) and Zn (II) by the dendrimer ligand and using polymer-supported ultrafiltration.[12]

3.3.3   METAL-CONTAINING NANOPARTICLES

Metal-containing nanoparticles such as metal and metal oxide nanoparticles have been studied extensively for water treatment applications. Most notable among the metal nanoparticles are the noble metals like silver and gold which are well-known as biocides; and nano-zerovalent iron (NZVI) used for treatment of water containing pesticides. Among metal oxides, the oxides of iron, aluminum, zinc, magnesium and titanium have been used in groundwater purification.

3.3.3.1   Zerovalent Iron (ZVI)

NZVI has applications in the removal of halogenated organics, arsenic, nitrate and heavy metals. Nanoparticles offer very high flexibility for both in-situ and ex-situ remediations. They can be anchored onto a solid matrix such as carbon, zeolite or membrane for enhanced water treatment. ZVI removes aqueous contaminants by reductive dechlorination in the case of chlorinated solvents, or by reducing to an insoluble form in the case of aqueous metal ions.

Supported zerovalent iron nanoparticles were used for separation and immobilization of Cr (VI) and Pb (II) from aqueous solution by reduction of chromium to Cr (III) and Pb to Pb (0).[13] Carboxymethyl cellulose (CMC) stabilized ZVI nanoparticles were used to reduce Cr (VI) in aqueous media through batch and continuous flow column. A stabilized ZVI nanoparticle was found to be more effective than a non-stabilized one for the removal of Cr (VI). In the batch experiments, the reduction of Cr (VI) was improved from 24% to 90% as the dosage of ZVI increased from 0.04 to 0.12 g/L. Nanopowder of ZVI was used for the removal of nitrate in water.[14] Studies have been carried out on the removal of As(V) by synthetic nanoscale zerovalent iron (NZVI). The effect of pH, adsorption kinetics, the sorption mechanism and anionic effects were investigated.[15] Studies on degradation of perchlorate in water indicated that stabilized ZVI nanoparticles could increase the perchlorate reduction rate by 53% in saline water (with concentration of NaCl up to 6% w/w).[16] ZVI nanoparticles and a commercial form of ZVI powder with different mesh sizes were used for the dechlorination of p-chlorophenol from water. The nanoparticles were found to be more effective for the reduction process.[17]

3.3.3.2 Noble-Metal Nanoparticles

Studies have been reported on the reaction of noble-metal nanoparticles with widely used pesticides.[18] The noble-metal nanoparticles supported on alumina appear quite effective for the removal of pesticides from solution. Pesticides found in drinking water are organochlorine (e.g., simazine, lindane, atrazine, etc.) or organosulfur pesticides (e.g., triazophos, quinalphos, etc.) or contain nitrogen-based functional groups (e.g., carbaryl, carbofuran, etc.). The chemistry of supported noble-metal nanoparticles can comfortably be utilized for the complete removal of such pesticides from drinking water. The complete removal of a wide variety of pesticides makes the chemistry of supported noble-metal nanoparticles unique and very promising for drinking water purification. The antimicrobial effects of silver, in zerovalent and ionic form, have also been investigated in detail, including the chemistry behind the biocidal activity of silver nanoparticles.[19-21]

3.3.3.3 Metal Oxide Nanoparticles

The major reasons for the use of metal oxide nanoparticles for water purification are high surface area for adsorption, mesoporous structure, surface charge, stability and low solubility in water. Studies have been carried out on different types of metal oxide nanoparticles. Nanoparticles activated by light, such as the large band-gap semiconductors like zinc oxide (ZnO) and titanium dioxide (TiO_2), have the ability to remove organic contaminants from wastewater. These nanoparticles have the advantages of availability, cost-effectiveness and low toxicity. The effect of size, fabrication method and morphology of ZnO nanoparticles on the decomposition of methyl orange (MO) has been studied. It was found that the preparation method is an important step and ZnO nanoparticles of 50 nm diameter range synthesized by thermal evaporation provided the highest activity of photocatalyst.[22] Alumina nanoparticles have been used for the removal of heavy metals from drinking water. The removal mechanism is based on the metal ion-induced flocculation of negatively charged alumina nanoparticles (alkaline pH conditions). Alumina is used as support for heterogeneous catalysis.[18] Titanium dioxide (TiO_2) is widely used in water treatment applications because of its photocatalytic properties. The rapid recombination of photo-generated electron hole pairs and the non-selectivity of the system are the main issues limiting the application of photocatalysis processes. Specific chelating agents such as arginine, lauryl sulfate and salicylic acid can modify the surface properties of nanocrystal TiO_2 and inhibit rapid recombination of photo-generated electron hole pairs. Fine TiO_2 particles have shown better efficiency than immobilized catalysts, but complete separation and recycling of fine particles (less than 0.5 μm) from treated water are very expensive. This issue can be solved by fixing carbon black modified nano-TiO_2 (CB-TiO_2) on aluminum sheet as a support. Carbon grain coated with activated nano-TiO_2 (TiO_2/AC) in the size range of 20–40 nm has been prepared and used for the photodegradation of MO dyestuff in aqueous solution under UV irradiation. The recovery of nanoparticles is easy when the particles are immobilized.[23] Silver nanoparticles doped TiO_2 were prepared and used for the photodegradation of direct azodyes.[24]

3.3.4 CARBON NANOTUBES (CNTs)

A carbon nanotube is a one-atom thick sheet of graphite (called graphene) rolled up into a seamless cylinder with a diameter of the order of a nanometer and capped at both ends by hemispheres of fullerene. This results in a nanostructure with length-to-diameter ratio of more than 10,000. CNTs can be categorized by their structures as single-walled nanotubes (SWNT) or multiwalled nanotubes (MWNT). As shown in Figure 3.1, an MWCNT is composed of a concentric arrangement of several cylinders. The high curvature of the graphene sheets increases the total energy of the tubules per carbon atom, but this is more than offset by a lowering of the energy because of the absence of dangling bonds at the edges of the graphene sheets. Such cylindrical carbon molecules have novel properties which make them potentially useful in a wide variety of applications in water treatment and healthcare.

The as-grown or acidified CNTs have good potential as adsorbing media for removal of contaminants from water. The as-grown CNTs have defect sites that originate during synthesis, which make them a suitable adsorbent for uptake of contaminants. However, such CNTs do not have any selectivity towards uptake of specific contaminants. The surface area of as-grown CNTs lies in the range of 50–100 m²/g. The surface areas of activated charcoal or activated alumina lie in the range of 100–200 m²/g, and these are extensively used in water decontamination because of their cost-effectiveness and easy availability.

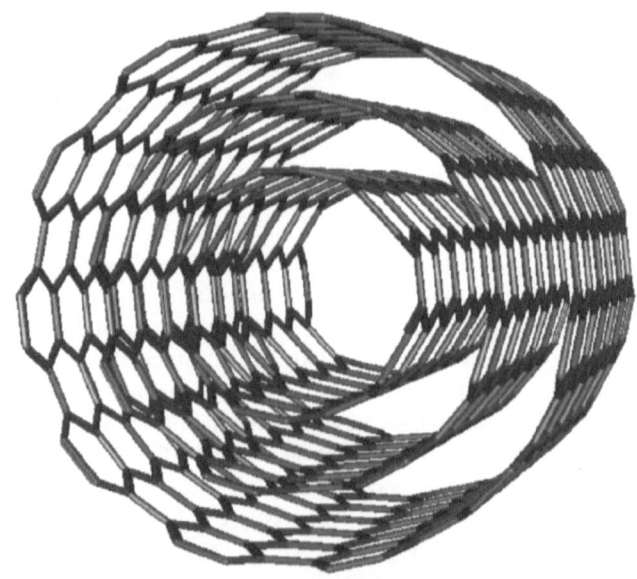

FIGURE 3.1   Multiwalled carbon nanotube (MWCNT).

Ceria supported on CNTs (CeO$_2$-CNTs) adsorbent has been reported for the removal of arsenate from water.[25] Under natural pH conditions, an increase from 0 to 10 mg/L in the concentration of Ca (II) and Mg (II) results in an increase from 10 to 81.9 and 78.8 mg/g in the amount of As (V) adsorbed, respectively. The efficient regeneration of the loaded adsorbent was carried out and the adsorption mechanism was suggested. The use of aligned carbon nanotubes (ACNTs) for the removal of fluoride from water has been reported.[26] The highest adsorption capacity of ACNTs occurs at neutral pH and reaches 4.5 mg/g at equilibrium fluoride concentration of 15 mg/L. Studies have been carried out on the removal of cadmium (II) with as-grown and surface-oxidized CNTs.[27] Cadmium (II) adsorption capacity for three kinds of oxidized CNTs was found to be more than for the as-grown CNTs due to the functional groups introduced by oxidation. The cadmium (II) adsorption capacity of the as-grown CNTs is only 1.1 mg/g, while it reaches 2.6, 5.1, and 11.0 mg/g for the H$_2$O$_2$, HNO$_3$ and KMnO$_4$-oxidized CNTs, respectively, at the cadmium (II) equilibrium concentration of 4 mg/L. Adsorption of cadmium (II) by CNTs was found to be strongly pH-dependent. CNTs show high adsorption capability and high adsorption efficiency for lead removal from contaminated water.[28] However, the adsorption was significantly influenced by the pH value of the solution and the surface characteristics of the nanotube, which can be controlled by their treatment processing.

A special case of chemical vapor deposition (CVD) is known as spray pyrolysis, in which liquid precursors like a mixture of benzene and ferrocene can be used. A self-standing tubular macro-geometry of aligned carbon nanotubes with high surface area (about 90 m^2/g) was developed using spray pyrolysis of benzene and ferrocene mixture.[29] The experimental set-up consists of a spray system, a container for the liquid precursor and a quartz tube. The spray system has an inner nozzle and an outer nozzle. The inner nozzle carries the liquid precursor and the outer nozzle carries the nitrogen gas. The spray system is attached to a quartz tube kept inside a resistive furnace. Optimized conditions of composition, flow rate of liquid precursor solution, carrier gas flow rate, diameter of the spray system nozzles and temperature of growth help in the synthesis of CNTs with particular length, diameter and alignment. Of the different methods of CNT synthesis such as arc discharge, laser ablation and CVD, CVD has good potential for scale-up. Figure 3.2 shows the micrograph of the self-standing microtube made up of aligned carbon nanotubes.

Figure 3.3 gives the scanning electron microscope (SEM) image of the self-standing tube, showing the alignment of CNTs in the radial direction. It shows the highly aligned CNTs and also the thickness of the cylinder. This has the advantages of a membrane where water purification can take place based on the size exclusion principle.

High aspect ratio, molecularly smooth hydrophobic graphite walls and nanoscale inner diameters of carbon nanotubes contribute to the ultra-efficient transport of water and gas through these narrow molecular pipes. Water and gas molecules move through nanotube pores at a speed that is orders of magnitude faster than through other pores of comparable size. It is noteworthy that CNTs possess many attributes that make them a particularly suitable nanostructured material for water purification. The challenge lies in development and implementation of CNTs in a device that can utilize the potential benefits.

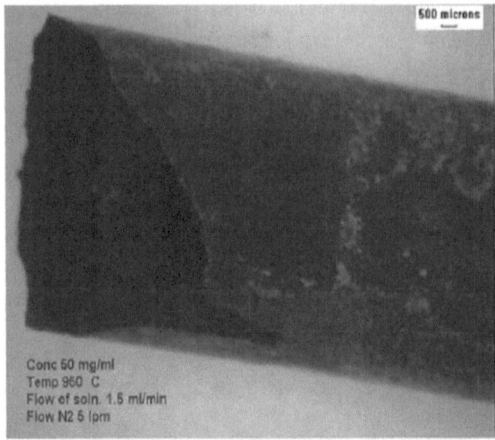

FIGURE 3.2   Micrograph of self-standing tube made up of aligned carbon nanotubes.

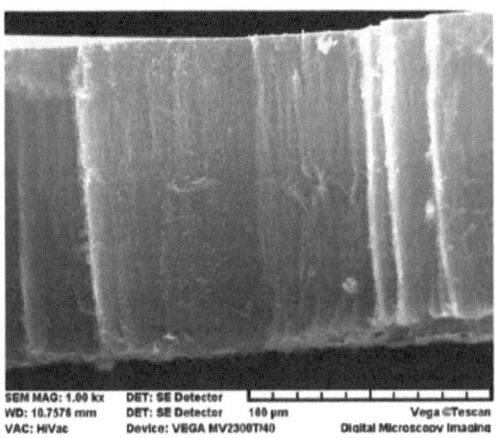

FIGURE 3.3   SEM image of the self-standing tube.

3.3.5   OTHER NANOMATERIALS

Clays are alumino-silicates with a planar silicate structure. There are three main categories of clay: kaolinite, montmorillonite-smectite and illite. The first two have
been widely investigated.[30] The structure contains silicate sheets (Si_2O_5) bonded to
aluminum oxide/ hydroxide layers ($Al_2(OH)_4$) called gibbsite layers. The primary
structural unit of this group is a layer composed of one octahedral sheet with one
tetrahedral sheet. Kaolinite is 1:1 clay mineral. The condensation of two sheets hap-
pens by coordination of an oxygen atom with one silicon atom in the tetrahedral
sheet and two aluminum atoms in the octahedral sheet. Clays also undergo exchange

interactions of adsorbed ions with the outside. Although clays have several advantages, they have one major disadvantage with respect to lack of permanent porosity. The clays swell and increase the space between their layers to accommodate the adsorbed water and ionic species. Different kinds of TiO_2 pillared clays from different raw clays and the adsorption and photocatalytic decomposition performance were evaluated.[31, 32] It was found that surface hydrophobicity of pillared clays (especially TiO_2) largely varied with the host clay. Since the TiO_2 particles in the pillared clays were too small to form a crystal phase, they presented poor photocatalytic activity. Nanocomposite of iron oxide and silicate was synthesized for degradation of azo-dye orange (II).[33] A new class of nanosized large porous titanium silicate (ETAS-10) and aluminum-substituted ETAS-10 with different Al_2O_3/TiO_2 ratios were successfully synthesized and used for the removal of heavy metals, such as Pb (II) and Cd (II).[34]

Studies have been carried out on magnetic nanoparticles as adsorbents and nanocatalysts for water treatment. One of the major applications of magnetic particles is in the area of magnetic separation, through which it is possible to separate a specific substance from a mixture of different other substances. It is called magnetically assisted chemical separation (MACS). Studies have been attempted on combining nanoparticle adsorption and magnetic separation for the removal and recovery of Cr(VI) from wastewater. Magnetic chitosan nano-adsorbent appears to be quite efficient for the removal of Co(II) ions at a pH range of 3–7 and a temperature range of 20–45°C.[35] Studies on the removal of nickel ions from aqueous solution using magnetic alginate microcapsules were also conducted.[36] Studies were carried out on the effect of particle size on the adsorption and desorption of arsenic from the point of view of arsenic removal.[37] Different kinds of magnetic nanoparticles were also employed for the removal of organic pollutants, such as sorption of methylene blue on polycyclic acid-bound iron oxide from an aqueous solution.[38]

3.4 SYNTHESIS OF NANOMATERIALS

There are a number of techniques for synthesizing different nanomaterials.[39] Nanoparticles can be produced from larger structures with top-down approaches such as using ultrafine grinders, lasers and vaporization followed by cooling. For complex particles, nanostructures are generally synthesized using a bottom-up approach such as arranging molecules to form complex structures with new and unique properties. Different synthesis routes are given below:

- Layer-by-layer deposition[40]
- Self-assembly[41, 42]
- Gas phase synthesis and sol-gel processing[43–45]
- Crystallization[46]
- Microbial synthesis[47, 48]
- Other methods such as sonochemical processing, cavitation processing, micro-emulsion processing and high-energy ball milling

3.5 NANOTECHNOLOGY: HEALTH, SAFETY AND ENVIRONMENT

Nanotechnology provides unprecedented technological solutions to water contamination issues enabling economic growth through efficient and durable products and new markets. However, the applicability of nanotechnology should be assessed after due consideration of the probable health and environmental risks.

Nanotechnology has the potential to offer clean synthesis and processing of nanoscale materials, reducing consumption of raw materials and natural resources such as water and energy. It is important to carry out a proper life-cycle analysis of the nanomaterials using validated nano-specific risk assessment methodologies.

Although there are only a limited number of products on the market that contain engineered nanomaterials, the pace of nanotechnology development may lead to a bigger market with nano-based products. In such a case, it is important to carry out an assessment of the associated health and environmental risks with respect to the following characteristics of nanoparticles:

- Size of particles
- High reactivity and conductivity
- Routes of exposure such as inhaling, ingestion, crossing the blood–brain barrier[49]

Nanomaterials constitute a new generation of toxic chemicals. As particle size decreases, the production of free radicals and toxicity increases in many nanomaterials. Studies show that some nanomaterials have the potential to impair the function or reproduction cycles of earthworms which play an important role in the nutrient cycling that underpins ecosystem function.[50] Studies have demonstrated that certain carbon nanotubes, when introduced into the lungs of rodents, cause inflammation, granuloma development, fibrosis, artery plaque responsible for heart attacks and DNA damage.[51–53] Two independent studies have shown that some carbon nanotubes can also cause the onset of mesothelioma, a cancer previously thought to be only associated with asbestos exposure.[54, 55] Very little is known about the safety risks of engineered nanomaterials. Their unique properties, such as increased reactivity and electrical conductivity, give rise to safety concerns with respect to causing fire or explosion. Because nanoparticles behave differently from larger particles, questions arise as to whether water supplies may be polluted or crops damaged during processes that release these particles into the air, soil or water. Laboratory researchers and production staff may be exposed to health and safety risks during the preparation and manufacture of nanomaterials. People in these occupations must be aware of the potential hazards of using materials that have unique properties, and they must take measures to mitigate the risks. However, their activities are contained and generally do not pose a threat to the general public or to the environment.

Owing to its highly interdisciplinary nature, nanotechnology is regarded as a valuable technology augmenting existing technologies in the field of water purification and healthcare. Bearing in mind the unique and at times unknown behavior of nanomaterials in the environment, nanotechnology may pose challenges to existing waste management systems. Knowledge of the mobility, persistence and bioaccumulation potential in the environment is very limited. Hence risk assessment of the possible impact of nano-wastes is critical and needs to be carried out.

Nanoparticles represent an entirely new risk. A proper and comprehensive risk and life-cycle analysis encompassing production, application and waste management strategies of nano-materials is urgently required for the successful commercialization of a technology with proven societal benefits. An extensive analysis of the risk can form the basis for government and international regulation.

3.6 DOMESTIC WATER PURIFICATION

Some of the major contaminants in drinking water and commonly used treatment solutions are given in Table 3.1. Treating water at the household level is quite an effective means of preventing waterborne disease. Simple and affordable interventions are capable of dramatically improving the microbial quality of household stored water and reducing the risks of waterborne diseases for ordinary people. Promoting household water treatment and safe storage (HWTS) has potential to help vulnerable populations to take charge of their own water quality by providing them with the knowledge and tools to treat their own drinking water. Household-level approaches to drinking water treatment and safe storage are also commonly referred to as managing water at the point of use (POU). Household water management is commonly used. There are a variety of treatment procedures, for example, with chlorine or other chemical disinfectants, sunlight or UV lamps, various filters or flocculation-disinfection formulations.[56]

TABLE 3.1
Water Contaminants and Commonly Used Treatment Solutions

S. No.	Contaminant	Possible Cause	Commonly Used Treatment Solution
1	Arsenic	Naturally occurring in water	Reverse osmosis, ion exchange, adsorption methods
2	Bacteria	Sewage, surface run-off, wells	Chlorination, ozonation, ultraviolet treatment, ultrafiltration
3	Lead	Industrial effluent, lead pipes	Reverse osmosis, anion exchange
4	Nitrate	Septic system, industrial effluent	Reverse osmosis, anion exchange, distillation
5	Pesticides and organic chemicals	Use of pesticides, chemicals near water source	Activated carbon filter, reverse osmosis, distillation
6	Odor, color, taste	Variety of sources	Activated carbon, ion exchange, chlorination
7	Turbidity	Fine sand, clay, other particles	Microfiltration
8	Hardness	Naturally occurring minerals calcium and magnesium in water	Ion exchange, water softener

TABLE 3.2
Nanoparticles for Removal of Contaminants from Water

S. No.	Nanoparticles	Contaminants Removed
1	Silver-magnesium oxide (Ag-MgO) nanoparticles	Bacteria
2	Zerovalent iron/carbon fiber	Organics
3	TiO$_2$/CNT/bimetallic nanoparticle	Metal ions

Table 3.2 gives some of the nanoparticles with potential to remove certain type of contamination in raw water. Ag-MgO nanoparticles are quite useful in removing bacteria from water due to its biocide characteristics and other added advantages of nanosize. Zerovalent iron/carbon-fiber nanoparticles are good for removing organic contamination in water, whereas TiO$_2$, CNT and bimetallic nanoparticles remove metal ions from raw water.

3.6.1 Sustainability of Water Purification Technology

There are several types of water filtration units that have the capability to purify contaminated water. However, many of them are too expensive for resource-poor people. Affordability is an important criterion for sustainability. Technologically advanced filters may not find application potential in marginalized countries without the capability to sustain them. Hence, cost-effective water purification methods are needed to have an impact on the global clean water shortage. The United Nations (UN) and World Health Organization (WHO) are currently pushing the industry to develop sustainable solutions to empower resource-poor users with the ability to purify their own water in a cost-effective and environmentally friendly way. These sustainable technologies are innovative and simple, incorporating a combination of basic science and local materials to create usable and efficient filters.

Sustainable technology is defined as technology that can be made at an affordable price by ordinary people using local materials to do useful work in ways that do the least possible harm to both humans and the environment.[57] The following guidelines are useful for the development of sustainable technology:

- Limit non-renewable energy consumption
- Lessen environmental impact
- Use readily available and easy-to-manufacture materials
- Safe and efficient manufacturing processes
- Robustness
- Scalability
- Easy to operate and maintain the product
- Mindful of cultural principles, practices or customs

3.6.2 CHALLENGES OF INTEGRATED NANO-BASED WATER PURIFICATION SYSTEMS

There are a number of different types of challenges associated with the development of successful nano-based technologies and products.

- Availability of nanomaterials
- Integration of nanomaterials into water purification systems
- Societal implications because of health and environment risks

3.6.2.1 Availability of Nanomaterials

The databank of different routes of production and usage of nanomaterials strongly suggests that there would be no shortage of nanomaterials either at present or in the near future. As demand rises, production would automatically try to cope with it. Studies carried out by Freedonia Group provide data on the demand for nanomaterials in the US for the years 2000 and 2003.[58] It also includes demand for the years 2008, 2013 and 2020 for materials (such as metal oxides, clays, metals, polymers, nanotubes, dendrimers, etc.) and applications (like abrasives, biocides, pharmaceutical fillers, catalysts, structural materials, etc.). These forecasts anticipate that most nanomaterials are nanoscale versions of established materials such as silica, titanium dioxide, clays and metal powders. Demand will increase as these nanomaterials become key components of several application industries in the field of water purification, electronics and healthcare and larger quantities of carbon nanotubes, fullerenes and dendrimers will also be available.

3.6.2.2 Integration of Nanomaterials into Water Purification Systems

This is a serious challenge with respect to potential adverse impact on health and environment that has to be overcome if the potential benefits of nanomaterials are to be exploited. Most nanomaterial usage is normally based on applications identified with fine powders. As we saw in the section on sustainability, the nanoparticles need to be impregnated in a chemically compatible and physically suitable host matrix so that a robust device can emerge with the potential for global applicability. The embedment of nanopowders in a host matrix has the following advantages:

- The agglomeration of the nanoparticles can be minimized, enabling better utilization of their desirable properties.
- The rate of reaction or sorption is enhanced because support can provide a high-concentration environment of target contaminant species around the loaded nanoparticles by adsorption.
- The contaminants react with the nano-surface or are oxidized on it. The resulting toxic intermediates can also get adsorbed on the support and as a result, they are not released into the atmosphere to cause secondary pollution.
- The sorption capability of the nanoparticle-support hybrid can be maintained for a long time.
- The recovery of nanoparticles from the host matrix is easy, leading to improved reusability.
- The possibility of secondary contamination of purified product water with nanomaterials can be minimized.

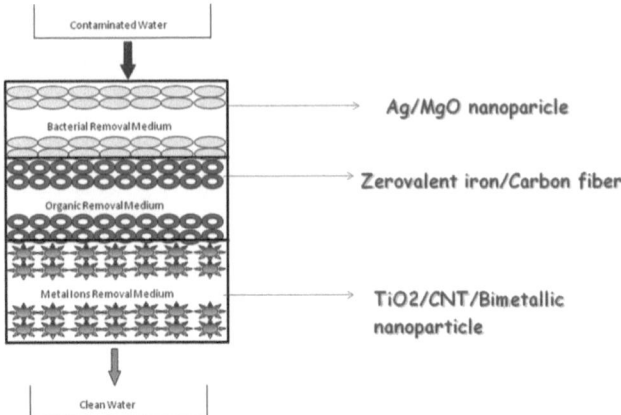

FIGURE 3.4 Composite nanomaterial packed-bed reactor for water purification.

Another important property that can be incorporated in a hybrid matrix is the option of treating contaminants of different kinds by a single system instead of dealing with each contaminant sequentially. Here the challenge will be to develop a cost-effective and environmentally acceptable separation and reactive medium that can be deployed in composite packed-bed reactors (Figure 3.4) for purification of water contaminated by (i) metal ions, (ii) organic solutes and (iii) bacteria.[59]

A membrane is a strong candidate host matrix for the incorporation of nanomaterials. It is used for a wide range of applications. The main property of a membrane is its ability to control the permeation rate of a chemical species. In separation applications, the objective is to allow one of the components of the feed stream to permeate through the membrane, while not allowing permeation of other components. A membrane separation process basically separates feed stream into two streams known as the permeate stream and the concentrate. The permeate stream is the stream containing constituents that have passed through the membrane, whereas the concentrate stream contains those that have been rejected by the membrane.

3.6.2.3   Societal Implications

As the impact of nanomaterials on human beings or the environment is not yet fully understood, further studies and extensive data are needed to speed up development and use of nanotechnology. The adverse environmental impact and toxicity of the material are the main issues regarding system design and material selection for water purification. Systematic investigations are required into the hydrolytic, oxidative, photochemical and biological stability of nanomaterials (e.g., dendrimers, carbonaceous nanoparticles, metal oxides, etc.) in natural and engineered environmental systems. Assessment of the risk of using nanomaterials presents unique challenges due to the paucity of published data. As with any new technology that offers significant benefits to humankind, there are also risks of adverse and unintended consequences with nanotechnology. Interdisciplinary discussions of the ethical and social dimensions of nanotechnology must be encouraged. Nanotechnology presents opportunities to integrate science and technology with social science and humanities. Education must provide mechanisms for updating scientists and engineers on new technologies as well as helping organize intelligent debates about the societal effects of nanotechnology.

REFERENCES

1. Whitesides, G.M. 2003. The "right" size in nanobiotechnology. *Nat. Biotechnol.* 21: 1161–1165.
2. Pradeep, T. and Anshup. 2009. Noble metal nanoparticles for water purification: A critical review. *Thin Solid Films* 517: 6441–6478.
3. Tratnyek, P.G. and Johnson, R.L. 2006. Nanotechnologies for environmental clean-up. *Nano Today* 1: 44–48.
4. Elizalde-Gonzalez, M.P., Mattusch, J., and Wennrichet, R. 2001. Application of natural zeolites for preconcentration of arsenic species in water samples. *J. Environ. Monit.* 3: 22–26.
5. Alvarez, A.E., Sanchez, A.G., and Querol, X. 2003. Purification of metal electroplating waste waters using zeolites. *Water Res.* 37: 4855–4862.
6. Wang, J. 2004. Characterizing the metal adsorption capability of a Class F coal fly ash. *Environ. Sci. Technol.* 38: 6710–6711.
7. Jeong, B.H., Hoek, M.V., Han, Y., Subramani, A., Huang, X., Hurwitz, G., Ghosh, A.K., and Jawor, A. 2007. Interfacial polymerization of thin film nanocomposites: A new concept for reverse osmosis membranes. *J. Membr. Sci.* 294: 1–7.
8. Bosman, A.W., Janssen, H.M., and Meijer, E.W. 1999. About dendrimers: Structure, physical properties and applications. *Chem. Rev.* 99: 1665–1688.
9. Tomalia, D.A. and Majoros, I. 2003. Dendrimeric supramolecular and supramacromolecular assemblies. *J. Macro Sci.* 43: 411–477.
10. Diallo, M.S., Christie, S., Swaminathan, P., Johnson Jr, J.H., and Goddard, W.A. 2005. Dendrimer enhanced ultrafiltration. 1. Recovery of Cu(II) from aqueous solutions using PAMAM dendrimers with ethylene diamine core and terminal NH_2 groups. *Environ. Sci. Technol.* 39: 1366.
11. Arkas, M., Tsiourvas, D., and Paleos, C.M. 2003. Functional dendrimeric "nanosponges" for the removal of polycyclic aromatic hydrocarbons from water. *Chem. Mater.* 14: 2844–2847.
12. Rether, A. and Schuster, M. 2003. Selective separation and recovery of heavy metal ions using water-soluble *N*-benzoylthiourea modified PAMAM polymers. *React. Funct. Polym.* 57: 13–21.
13. Ponder, S.M., Darab, J.G., and Mallouk, T.E. 2000. Remediation of Cr(VI) and Pb(II) aqueous solutions using supported, nanoscale zero-valent iron. *Environ. Sci. Technol.* 34: 2564–2569.
14. Choe, S., Chang, Y.Y., Hwang, K.Y., and Khim, J. 2000. Kinetics of reductive denitrification by nanoscale zero-valent iron. *Chemosphere* 41: 1307–1311.
15. Kanel, S.R., Greneche, J.M., and Choi, H. 2006. Arsenic(V) removal from groundwater using nanoscale zero-valent iron as a colloidal reactive barrier material. *Environ. Sci. Technol.* 40: 2045–2050.
16. Xiong, Z., Zhao, D., and Pan, G. 2007. Rapid and complete destruction of perchlorate in water and ion-exchange brine using stabilized zero-valent iron nanoparticles. *Water Res.* 41: 3497–3505.
17. Cheng, R., Wang, J.L., and Zhang, W.X. 2007. Comparison of reductive dechlorination of *p*-chlorophenol using FeO and nanosized FeO. *J. Hazard. Mater.*, 144: 334–339.
18. Nair, A.S. and Pradeep, T. 2007. Extraction of chlorpyrifos and malathion from water by metal nanoparticles. *J. Nanosci. Nanotechnol.* 7: 1871–1877.

19. Jain, P. and Pradeep, T. 2005. Potential of silver nanoparticle-coated polyurethane foam as an antibacterial water filter. *Biotechnol. Bioeng.* 90: 59–63.

20. Sambhy, V., MacBride, M.M., Peterson, B.R., and Sen, A. 2006. Silver bromide nanoparticle/polymer composites: Dual action tunable antimicrobial materials. *J. Am. Chem. Soc.* 128: 9798–9808.

21. Silver, S. 2003. Bacterial silver resistance: Molecular biology and uses and misuses of silver compounds. *FEMS Microbiol. Rev.* 27: 341–353.

22. Wang, H., Xie, C., Zhang, W., Cai, S., Yang, Z., and Gui, Y. 2007. Comparison of dye degradation efficiency using ZnO powders with various size scales. *J. Hazard. Mater.* 141: 645–652.

23. Li, Y., Li, X., Li, J., and Yin, J. 2006. Photocatalytic degradation of methyl orange by TiO$_2$-coated activated carbon and kinetic study. *Water Res.* 40: 1119–1126.

24. Sobana, N., Muruganadham, M., and Swaminathan, M. 2006. Nano-Ag particles doped TiO$_2$ for efficient photodegradation of direct azo dyes. *J. Mol. Catal. A: Chem.* 258: 124–132.

25. Peng, X., Luan, Z., Ding, J., Di, Z., Li, Y., and Tian, B. 2005. Ceria nanoparticles supported on carbon nanotubes for the removal of arsenate from water. *Mater. Lett.* 59: 399–403.

26. Li, Y., Wang, S., Zhang, X., Wei, J., Xu, C., Luan, Z., and Wu, D. 2003. Adsorption of fluoride from water by aligned carbon nanotubes. *Mater. Res. Bull.* 38: 469–476.

27. Li, Y., Wang, S., Luan, Z., Ding, J., Xu, C., and Wu, D. 2003. Adsorption of Cd(II) from aqueous solution by surface oxidized carbon nanotubes. *Carbon* 41: 1057–1062.

28. Li, Y., Wang, S., Wei, J., Zhang, X., Xu, C., Luan, Z., Wu, D., and Wei, B. 2002. Lead adsorption on carbon nanotubes. *Chem. Phys. Lett.* 357: 263–266.

29. Dasgupta, K., Kar, S., Venugopalan, R., Bindal, R.C., Prabhakar, S., Tewari, P.K., Bhattacharya, S., Gupta, S.K., and Sathiyamoorthy, D. 2008. Self-standing geometry of aligned carbon nanotubes with high surface area, *Mater. Lett.* 62: 1989–1992.

30. Pradeep, T. and Anshup. 2009. Detection and extraction of pesticides from drinking water using nanotechnologies. In: Savage, N., Diallo, M., Duncan, J., Street, A., and Sustich, R. (Eds.), *Nanotechnology Application for Clean Water.* New York: William Andrew Publications.

31. Ding, Z., Zhu, H.Y., Lu, G.K., and Greenfield, P.F. 1999. Photocatalytic properties of Titania pillared case by different drying methods. *J. Colloid Inter. Sci.* 209: 193–199.

32. Ooka, C., Yoshida, H., Suzuki, K., and Hattori, T. 2004. Highly hydrophobic TiO$_2$ pillared clay for photocatalytic degradation of organic compounds in water. *Micro. Meso. Mater.* 67: 143–150.

33. Feng, J., Hu, X., Yue, P.L., Zhu, H.Y., and Lu, G.Q. 2003. Degradation of azo-dye orange II by a photo-assisted fenton reaction using a novel composite of iron oxide and silicate nanoparticles as catalyst. *Ind. Eng. Chem. Res.* 42: 2058–2066.

34. Choi, J.H., Kim, S.D., Noh, S.H., Oh, S.J., and Kim, W.J. 2006. Adsorption behaviors of nano-sized ETS-10 1nd Al-substituted-ETS-10 in removing heavy metal ions. *Micro. Meso. Mater.* 87: 163–169.

35. Chang, Y.C., Chang, S.W., and Chen, D.H. 2006. Magnetic chitosan magnetic particles: Studies on chitosan binding and adsorption of Co(II) ions. *React. Funct. Polym.* 66: 335–341.

36. Ngomsik, A.F., Bee, A., Siaugue, J.M., Cabuil, V., and Cote, G. 2006. Nickel adsorption by magnetic alginate microcapsules containing an extractant. *Water Res.* 40: 1848–1856.

37. Mayo, J.T., Yavuz, C., Yean, S., Cong, L., Shipley, H., Yu, W., Falkner, J., Kan, A., Tomson, M., and Colvin, V.L. 2007. The effect of nanocrystalline magnetic size on arsenic removal. *Sci. Technol. Adv. Mater.* 8: 71–75.

38. Mak, S.Y. and Chen, D.H. 2004. Fast adsorption of methylene blue on polyacrylic acid-bound iron oxide magnetic nano-particles. *Dyes Pigments* 61: 93–98.

39. Tiwari, D.K., Behari, J., and Sen, P. 2008. Application of nanoparticles in wastewater treatment. *World Appl. Sci. J.* 3: 417–433.

40. Phillips, K.S., Han, J.H., Martinez, M., Wang, Z.Z., Carter, D., and Cheng, Q. 2006. Nanoscale glassification of gold substrates for surface plasma resonance analysis of protein toxins with supported lipid membranes. *Anal. Chem.* 78: 596–603.

41. Graveland, B.J.F. and Kruif, C.G. 2006. Unique milk protein based nanotubes: Food and nanotechnology meet. *Trends Food Sci. Technol.* 17: 196–203.

42. Lorenceau, E., Utada, A.S., Link, D.R., Cristobal, G., Joanicot, M., and Weitz, D.A. 2005. Generation of polymersomes from double-emulsions. *Langmuir* 21: 9183–9186.

43. Siegel, R.W. 1991. Nanomaterials: Synthesis, properties and applications. *Ann. Rev. Mater. Sci.* 21: 559–578.

44. Siegel, R.W. 1994. Physics of new materials. In: Fujita, F.E. (Ed.), *Springer Series in Materials Science*, Vol. 27. Berlin: Springer.

45. Uyeda, R. 1991. Studies of ultrafine particles in Japan: Crystallography, methods of preparation and technological applications. *Prog. Mater. Sci.* 35: 1–96.

46. Boanini, E., Torricelli, P., Gazzano, M., Giardino, R., and Bigi, A. 2006. Nanocomposites of hydroxyapatite with aspartic acid and glutamic acid and their interaction with osteoblast-like cells. *Biomaterials* 27: 4428–4433.

47. Bhainsa, K.C. and Souza, S.F. 2006. Extracellular biosynthesis of silver nanoparticles using the fungus *Aspergillus fumigatus*. *Colloids Surf. B: Biointerfaces* 47: 160–164.

48. Bhattacharya, D. and Gupta, R.K. 2005. Nanotechnology and potential of microorganisms. *Crit. Rev. Biotechnol.* 25: 199–204.

49. Oberdorster, G., Sharp, Z., Atudonrei, V., Elder, A., Gelein, R., Kreyling, W., and Cox, C. 2004. Translocation of inhaled ultrafine particles to the brain. *Inhalation Toxicol.* 16: 453–459.

50. Scott-Fordsmand, J., Krogh, P., Schaefer, M., and Johansen, A. 2008. The toxicity testing of double-walled nanotubes contaminated food to *Eisenia veneta* earthworms. *Ecotoxicol. Environ. Safety* 71: 616–619.

51. Donaldson, K., Aitken, R., Tran, L., Stone, V., Duffin, R., Forrest, G., and Alexander, A. 2006. Carbon nanotubes: A review of their properties in relation to pulmonary toxicology and workplace safety. *Toxicol. Sci.* 92: 5–22.

52. Lam, C.W., James, J., McCluskey, R., Arepalli, S., and Hunter, R. 2006. A review of carbon nanotube toxicity and assessment of potential occupational and environmental health risks. *Crit. Rev. Toxicol.* 36: 189–217.

53. Muller, J., Huaux, F., and Lison, D. 2006. Respiratory toxicity of carbon nanotubes: How worried should we be? *Carbon* 44: 1048–1056.

54. Poland, C., Duffin, R., Kinloch, I., Maynard, A., Wallace, O., Seaton, A., Stone, V., Brown, S., MacNee, S.W., and Donaldson, K. 2008. Carbon nanotubes introduced into the abdominal cavity of mice show asbestos like pathogenicity in a pilot study. *Nat. Nanotechnol.* 3: 423–428.

55. Takagi, A., Hirose, A., Nishimura, T., Fukumori, N., Ogata, A., Ohashi, N., Kitajima, S., and Kanno, J. 2008. Induction of mesothelioma in p53+/– mouse by intraperitoneal application of multi-wall carbon nanotube. *J. Toxicol. Sci.* 33: 105–116.

56. WHO. 2007. *Report on "Combating waterborne disease at the household level."* The International Network to Promote Household Water Treatment and Safe Storage.

57. Cunningham, W.P., Cunningham, M.A., and Saigo, B.W. 1999. *Environmental Science: A Global Concern* (7th ed.). New York: McGraw Hill.

58. Freedonia Group Inc. 2005. Nanomaterials—Market size, market share, market leaders, demand forecast, sales, company profiles, market research, industry trends. Available at: www.freedoniagroup.com

59. Savage, N. and Diallo, M.S. 2005. Nanomaterials and water purification: Opportunities and challenges. *J. Nanopart. Res.* 7: 5331–5342.

4 Nanocomposite Membranes in Water Treatment

4.1 INTRODUCTION

A conventional membrane has inherent limitations in water treatment and purification due to its low resistance to fouling and the trade-off that exists between permeability and selectivity. The nanocomposite membrane, made by combining polymeric materials with nanomaterials, is emerging as a potential alternative. It has good potential for applications in desalination, water purification, wastewater treatment, recycling and reuse of water, and can be developed to meet specific water treatment applications by tuning its structure and physico-chemical properties, including hydrophilicity, porosity, charge density, thermal and mechanical stability, and by introducing unique functionalities (e.g. antibacterial capabilities).

There are four types of nanocomposite membranes[1] (Fig. 4.1):

(i)   conventional nanocomposite;
(ii) thin-film nanocomposite (TFN);
(iii) thin-film composite (TFC) with nanocomposite substrate;
(iv) surface-located nanocomposite.

4.2 NANOCOMPOSITES

The nanocomposite membrane was originally developed for gas separation, in which zeolite was incorporated into polymer to enhance permeability as well as selectivity.[2, 3] A mixed-matrix nanocomposite membrane as a tunable water treatment membrane offers better selectivity, targeted functionalities and high thermal, chemical and mechanical stability. It has the advantage of the polymeric membrane's ease of preparation as well as high mechanical strength and the functional properties of inorganic materials.

Preparation of a nanocomposite membrane is normally via the phase inversion method. A nanofiller is dispersed in polymer solution prior to the phase inversion process, and prepared in either flat-sheet or hollow-fiber configuration (Figure 4.2). This type of membrane is mainly used in microfiltration or ultrafiltration due to its typical porous structure.

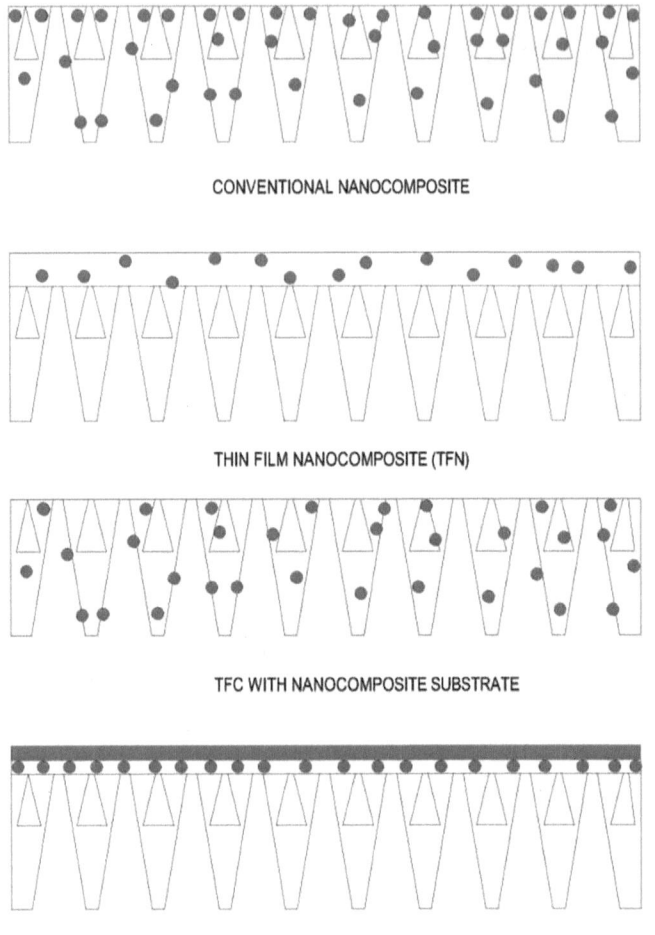

CONVENTIONAL NANOCOMPOSITE

THIN FILM NANOCOMPOSITE (TFN)

TFC WITH NANOCOMPOSITE SUBSTRATE

SURFACE LOCATED NANOCOMPOSITE

FIGURE 4.1 Types of nanocomposite membranes. (Reprinted from *J. Membr. Sci.*, 479, Yin, J., Deng, B. 2015. Polymer-matrix nanocomposite membranes for water treatment, 256–275, Copyright 2015, Figures 1, 2, 4 and 5.)

4.2.1 CARBON NANOTUBE REINFORCEMENT

Considerable research work has been carried out in the field of impregnation of CNTs in polymer host matrix as reinforcement for the incorporation of better properties like anti-biofouling and mechanical strength.[4–19] The CNT-embedded membrane has good potential in water treatment and purification. Multi-wall carbon nanotube/cellulose acetate (CNT/CA) nanocomposite membranes are prepared using the phase inversion method.[20, 21] Carbon nanotubes are functionalized by oxidation purification in a strong acidic medium to enhance their dispersion in the polymer matrix. CNTs are randomly oriented and uniformly dispersed within the

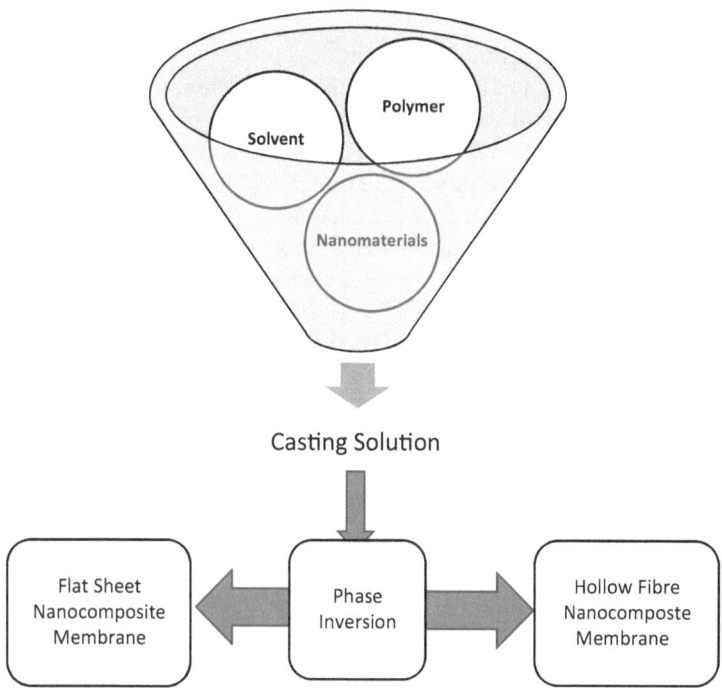

FIGURE 4.2 Preparation of nanocomposite membrane by phase inversion method.

membranes. It is observed that the increase in CNT content leads to a decrease in the number of macro voids. The pore size decreases with the increase in CNT content. Permeation rate improves significantly in membranes with low CNT content. Further addition of carbon nanotubes causes a reduction in permeation rate due to the decreased porosity and surface area. Membrane systems have been evaluated for the removal of the extractable organic fraction from oil sand process-affected water.[22] Experiments were performed using in-situ integrated membrane units consisting of low pressure-driven membrane and high pressure-driven membrane modules. The low pressure-driven membrane was prepared with a polysulfone phase inversion casting process. The high pressure-driven membrane was made by polyamide thin-film composite interfacial polymerization. The multi-wall carbon nanotubes were modified with strong acid to enhance dispersion in an organic solvent. Acid-modified multi-wall carbon nanotubes developed surface functional groups which increased their hydrophilicity, thus increasing the rejection of hydrophobic pollutants and significantly reducing membrane fouling.

The incorporation of carbon nanotubes disrupts the bulk polymer in the membrane and creates nanoscale cavities in the coating layer to increase the water flow.[23] Incorporation of carbon nanotubes improves the mechanical stability of the membrane. It has been observed that incorporation of multi-wall carbon nanotubes into electrospun polyvinyl alcohol (PVA) leads to improved mechanical strength and durability[24] as well as enhanced water flux due to frictionless nanotube channels. A functionalized multi-walled carbon nanotube (MWCNT) immobilized polyethyleneimine–poly(amide–imide) (PEI–PAI) hollow-fiber membrane was designed and prepared by phase inversion, followed by functionalized MWCNT immobilization by vacuum filtration before a chemical post-treatment using PEI to obtain a positively charged selective layer.[25] The resulting membrane gave 44% increase in water permeability without significant compromise on the salt permeability in forward osmosis.

It has been observed that the pure water flux can be increased in a carboxylated MWCNT blended polysulfone membrane prepared by phase inversion by increasing the MWCNT loading up to 1.5%.[26] Pure water flux decreases with further loading of multi-wall carbon nanotubes. The flux increases due to the hydrophilic surface and large surface pores resulting from the addition of MWCNTs. PEG6000 and multi-wall carbon nanotubes used for the preparation of chitosan porous membranes with 10 wt% multi-wall carbon nanotube loading gave several times higher water flux than pure chitosan membranes.[27] The higher water flux was observed due to the formation of multi-wall carbon nanotube nanochannels in chitosan pores. The tensile strength of the membrane increases with the addition of multi-wall carbon nanotubes.

Studies have shown that hydroxyl-functionalized multi-walled carbon nanotubes blended with polyacrylonitrile (PAN) to prepare ultrafiltration membranes using the phase inversion process leads to an increase in water flux of the membranes by 63% at 0.5 wt% loading of MWCNTs as compared to pure PAN membranes.[28] The water flux decreases as the concentration of MWCNTs is further increased; however, at 2 wt% loading it is still higher than for neat PAN membranes. The surface hydrophilicity of the membrane is enhanced by the addition of MWCNTs, as observed by contact angle measurements. The increased hydrophilicity leads to improved water flux. The tensile strength of the membrane with 2 wt% MWCNT loading is also greater than that of the neat membrane.

4.2.2 Metal Oxide Reinforcement

The incorporation of metal oxide nanomaterials (such as TiO_2,[29–63] SiO_2,[64–69] Al_2O_3,[70–72] Fe_3O_4,[73–77] MnO_2[78] and ZnO[79]) into polymers tunes their structure and physico-chemical properties, including hydrophilicity, porosity and charge density, and chemical, thermal and mechanical stability. It also introduces antifouling and photocatalytic characteristics into the membranes.

Fouling is one of the major limitations in membrane separation processes. One of the practical strategies used to overcome this limitation is the application of anti-biofouling membrane material. Nanocomposite membranes made of mesoporous silica particles[80] in polyethersulfone (PES) using the phase inversion method offer antifouling properties. A nanocomposite membrane with 2% mesoporous silica exhibits excellent hydrophilicity, water permeability and good antifouling performance. In addition, the introduction of the mesoporous silica particles improves the thermal stability of nanocomposite membranes. The protein adsorption on the membrane surface decreases significantly when the mesoporous silica content increases. The ultrafiltration experiments reveal that the incorporation of mesoporous silica particles in the membrane reduces membrane fouling, especially irreversible fouling. Higher mesoporous silica content (4%) results in particle agglomeration. The incorporation of spray-dried nanostructured silica granules into polysulfone nanocomposite membranes[81] gives significantly enhanced water permeability without sacrificing separation performance. Mechanical and chemical modifications of titania (TiO_2) nanoparticles reduce nanoparticle agglomeration and fouling in flat-sheet membranes. The migration of nanoparticles towards the outer layer occurs in mechanical modifications of particles, whereas the migration and size of agglomeration reduces significantly with mechanically and chemically modified titania particles.[82]

Higher thermal resistance and stiffness, and lower elasticity are observed in fibers made with chemically and mechanically modified particles. Enhancements in initial pure water flux due to lower intrinsic membrane resistance and bigger pore size is also observed, with no effect on salt rejection. Modification of a polyvinylidene fluoride (PVDF)/sulfonated polyethersulfone (SPES) blend membrane was attempted using titania nanoparticles.[83] In this case, sulfonation of polyethersulfone (PES) was carried out, then the PVDF/ SPES blend membrane was prepared with titania nanoparticles in the casting solution, using the phase inversion process induced by the immersion precipitation technique. About 4 wt% polyvinylpyrrolidone (PVP) was added in the casting solution as pore former. Addition of titania nanoparticles in the casting solution reduced the size of membrane pores in the surface and sublayers. Contact angle measurement demonstrates that the hydrophilicity of the modified membrane is enhanced by the addition of titania in the casting solution. The experimental results imply that the initial flux of the titania-entrapped PVDF/SPES membrane is lower than that of the neat PVDF/SPES membrane. The antifouling properties of the membrane are improved by changing the membrane surface from hydrophobic to hydrophilic after titania addition to the casting solution. The type and size of titania nanoparticles have an impact on the morphology, performance and fouling control aspects of mixed-matrix polyethersulfone nanofiltration membranes.[84] Surface hydrophilicity of the TiO_2 blended membrane improves due to improved water affinity in the membrane surface. The nanofiltration membrane made from the nano-sized titania particles enhanced the pure water flux and antifouling properties when tested for whey solution. The low concentration of titania nanoparticles leads to more reduction in biofouling because the aggregation of the nanoparticles does not take place at low amounts. Poly(phthalazine ether sulfone ketone)/titania (PPESK/TiO_2) organic–inorganic composite ultrafiltration (UF) membranes prepared by phase inversion perform better in treatments of high-temperature condensed water.[85] It is observed that the finger-like structure in the membrane sublayer is suppressed, and the sponge-like structure begins to be developed. The mechanical strength and thermal stability of the composite membrane are improved due to the existence of a hydrogen bond between titania and polymer. The permeate flux is also enhanced due to the improved hydrophilicity and porosity. Compared with a pure PPESK membrane, the composite membrane gives better antifouling properties, showing lower filtration resistance and better flux recovery during treatment of high-temperature condensed water.

4.2.3 Nanoclay Reinforcement

Polyvinylidene fluoride (PVDF) is a commonly used polymer in water treatment due to its chemical robustness. It has been observed that nanoclay reinforcement in PVDF improves the mechanical properties of flat-sheet membranes.[86, 87] Studies report on the wear properties of polymeric and nanocomposite materials, including clay nanocomposites;[88–90] however, this has not been extended to the membrane field. Improved abrasion resistance is observed when nanoclay is incorporated into nanocomposite materials and PVDF hollow-fiber membranes.[91] Polyvinylidene fluoride (PVDF)/nanoclay hollow-fiber membranes fabricated by nonsolvent-induced phase separation (NIPS) show significantly improved physical endurance. The incorporation of nanoclay shifts the crystalline phase of PVDF from α-phase to β-phase and improves the membrane structure as well as its mechanical properties in terms of stiffness and flexibility. Tensile strength increases from 3.8 MPa to 4.3 MPa with 5.08 wt% Cloisite-30B loading. Nanoclay gives reinforcement to nanocomposite membranes.[92–99]

4.2.4 Organic Materials Reinforcement

Polymers like polysulfone, polyethersulfone, polyacrylonitrile, cellulose acetate and ethylene vinyl alcohol are used to make conventional membranes. Organic materials like cyclodextrin,[100–103] polyaniline (PANI),[104–109] polypyrrole,[110] chitosan,[111] polyhedral oligomeric silsesquioxane[112] and semi-interpenetrating network polymeric nanoparticles[113] are used to increase hydrophilicity, and improve the adsorptive, anti-compaction and antifouling behavior of the resultant nanocomposite membranes.

4.2.5 Dendrimer Reinforcement

Dendrimers and their derivatives are substances with diverse analytical, biomedical and environmental applications[114–118] due to their unique molecular structure, easy functionalization and manipulation of their terminal groups.[119–124] Dendritic polymers consist of a multifunctional core, a high degree of repeated branching units and high density of surface functional groups.[125] A commercial polyamidoamine (PAMAM) dendrimer can be used in separation systems for recovery of heavy metals from aqueous solution by means of chelating agents in pollution remediation processes. In particular, an aqueous heavy metal solution treated with PAMAM before being passed through an ultrafiltration membrane has been proposed for water and soil remediation.[126–130] Raw PAMAM, aromatic PAMAM and PAMAM coated with polyethylene glycol can be used in the modification of polymeric reverse osmosis membranes for remediation of copper, nickel and chromium ions from wastewater.[131] Dendrimer inclusion improves the membrane's elastic behavior. This was demonstrated by embedding a diaminobutane-based poly(propyleneimine) dendrimer functionalized with sixteen thiol groups, DAB-3-(SH)16, in a swollen cellulosic support.[132] An increase of about 20% in Young's modulus is noted. A significant reduction in the permeation of toxic heavy metals (Cd^{2+}, Hg^{2+} and Pb^{2+}) is observed, indicating the possible application of dendrimer-modified membranes in electrochemical devices for water purification.

4.2.6 ZEOLITE REINFORCEMENT

Zeolites are microporous, alumina-silicate minerals commonly used as commercial adsorbents and catalysts. They belong to the family of microporous solids known as molecular sieves. Because the pore structure is of molecular dimensions, they have the ability to selectively sieve molecules based on size exclusion. The maximum size of the molecular or ionic species that can enter the pores of a zeolite is controlled by the dimensions of the channels. Zeolites are used as reinforcement materials in the polymeric host matrix[133–135] to increase hydrophilicity, and improve cross-linking and molecular sieving. Zeolite-PVDF nanocomposite membranes possess enhanced mechanical stability in terms of their tensile strength[136] due to the positive interaction between zeolite and the PVDF matrix, where zeolite nanoparticles act as a cross-linking agent for the polymeric chains and increase membrane rigidity. The cross-linking phenomenon is attributed to the hydrogen bonds between the polymer chains and hydroxyl groups T-OH on the zeolite surface, where T is silicon, aluminum or phosphorus.[137]

4.2.7 SILVER REINFORCEMENT

It is known that biofouling causes deteriorations in membrane performance, including flux deterioration, membrane degradation and increased operation and maintenance costs. To reduce biofouling, functional membranes containing biocides or antibacterial materials have attracted much interest. Silver has been widely explored because of its excellent biocidal properties.[138–140] Several studies have been reported on silver nanoparticles introduced into membrane materials such as polysulfone,[141, 142] polyethersulfone,[143–146] polyvinylidene fluoride,[147, 148] polyamide[149, 150] and chitosan.[151] The addition of silver nanoparticles improves membrane performance in terms of flux and fouling resistance, which is attributed to an increase of hydrophilicity or a change in membrane morphology. However, the chemically produced silver nanoparticles often have stability problems. The beneficial effects of adding particles are often mitigated by aggregation and poor compatibility with the polymeric matrix. Despite the improved durability of the silver-containing membrane and simultaneously reduced potential risks of releasing silver ions into the environment, silver nanoparticles still face some challenges with respect to membrane performance.

A silver nanoparticle (AgNP) embedded membrane has the potential to prevent biofouling permanently. However, this mechanism depends on the location of silver nanoparticles in the membrane matrix. The location of silver nanoparticles may change depending on the type of polymer and phase inversion parameters. Polysulfone (PS), polyethersulfone (PES) and cellulose acetate (CA) polymers are used for preparation of flat-sheet nanocomposite membranes. It is found that silver nanoparticles accumulate on the top and skin layers of PES and PS nanocomposite membranes. The interaction between silver nanoparticles and bacteria depends on the release of ionic silver from the silver-embedded membrane. The nanocomposite membranes that store silver nanoparticles at the surface have the best antibacterial properties. Antibacterial polysulfone ultrafiltration membranes are made using silver nanoparticles in the casting solution. Ultrafiltration mixed-matrix membranes are prepared using silver nanoparticles in polysulfone casting solution. The crystallinity of the membrane matrix decreases as the size of the silver nanoparticle decreases. Silver nanoparticles containing membranes showed high antibacterial properties, particularly with smaller silver nanoparticles.

The biogenic nanocomposite polyethersulfone membranes are prepared by adding different amounts of biogenic nanoparticle silver into the dope solution. The nanocomposite membranes are tested for physical properties with pure water permeability. They slightly increase the hydrophilicity of polyethersulfone (PES) membranes and improve water permeability. The protein adsorption on the membrane surface decreases significantly due to the increased hydrophilicity and improved smoothness of membrane surfaces. To obtain both organic antifouling and antibacterial properties, acrylamide is grafted onto a polyethersulfone (PES) hollow-fiber membrane, and silver nanoparticles (AgNPs) are formed within the acrylamide layer. The hydrophilicity of the membrane surface is improved by acrylamide grafting, leading to a reduction in membrane fouling. A bare PES membrane has no antibacterial activity and bacteria grow on the membrane surface, while a PES membrane containing silver nanoparticles indicates high antibacterial activity. Thus, polymer membranes containing silver nanoparticles within the acrylamide gel layer have high potential for organic antifouling and antibacterial applications.

A mixed-matrix porous polymeric membrane with antifouling properties is prepared by phase inversion from a quaternary system of polysulfone/N,N-dimethyl-formamide/poly-vinylpyrrolidone (PVP)/nanosilver (nAg). The effect of the composition of the casting mixture on membrane morphology, performance and antifouling properties is evaluated using microscopic, spectroscopic and surface characterization techniques. The incorporation of nanosilver into a casting mixture containing 5 wt% PVP introduces morphological changes in the membrane structure including increased pore size, suppression of macro voids and thinning of the skin layer. These changes enhance the separation and antifouling properties of membranes. Nanoparticles of silver, copper and silver–copper mixture have been impregnated into the polysulfone host matrix and the biofouling resistance behavior of each membrane surface has been examined.[152–154] The performance of the membranes has been evaluated for pure water permeability and solute rejection. The silver-impregnated membranes possess the highest resistance to biofouling. Nanomaterials such as copper[155–157] and selenium[158] have also been explored for their potential applications to antimicrobial membranes.

4.2.8 GRAPHENE OXIDE REINFORCEMENT

Graphene Oxide (GO) reinforcement is gaining interest due to its high surface area, outstanding electron transport and mechanical properties. When incorporated into the polymer matrix, the atomically thin carbon sheets can significantly improve physical properties of the host polymer at extremely low dope concentration. The graphene oxide has hydrophilic and pH-sensitive behavior. It exhibits a negative surface charge throughout the entire pH range. Due to different types of hydrophilic functional groups present on the surface of graphene oxide, it can take up water very easily. The water uptake increases as the degree of oxidation increases.[159] A graphene oxide dispersed polysulfone mixed-matrix membrane prepared using wet phase inversion exhibits improved salt rejection. The salt rejection shows an increasing trend with increase in the pH value. Polyvinylidene fluoride (PVDF)/graphene oxide (GO) ultrafiltration (UF) membranes prepared[160] by immersion precipitation phase inversion show large amounts of –OH groups due to the introduction of graphene oxide nanosheets which improve the surface hydrophilicity of the modified membrane. In permeation experiments, the water flux is improved after blending graphene oxide. The flux recovery ratio (FRR) and the fouling resistance suggest that PVDF/GO UF membranes have better antifouling properties than pure PVDF due to the changes in surface hydrophilicity and membrane morphologies. A graphene oxide nanosheet used as a nanofiller increases hydrophilicity and antifouling performance of a polymer-based membrane (such as PVDF), resulting in a high-performance ultrafiltration membrane with enhanced flux.[161]

Using the immersion phase inversion process, polyvinylidene fluoride and graphene oxide dissolved in N, N-dimethylacetamide (DMAc) to prepare an organic–inorganic blended ultrafiltration membrane,[162, 163] shows that the properties and structure of the blended membrane improve for 0.20 wt% graphene oxide addition in the casting solution. The permeability of the blended membrane increases by 96.4% with a slight change of retention. The tensile strength increases by 123%. The contact angle decreases from 79.2° to 60.7°, which implies that the antifouling ability of the membrane has increased.

A membrane bioreactor (MBR) is quite effective in terms of compactness and space requirement compared to conventional wastewater treatment processes.[164] However, the high cost of membrane material and maintenance due to membrane fouling restricts its application.[165] The bio-cake layer accumulated on the membrane surface is one of the major causes of membrane fouling in the MBR process.[166–168] It is observed that the PVDF/GO composite membrane demonstrates sustained permeability and lower cleaning frequency than that of the PVDF membrane.[169]

4.2.9 HYBRID MATERIALS REINFORCEMENT

Studies have been carried out on reinforcement of the membrane with hybrid materials to give better performance. Titania-coated multi-wall nanotubes reinforced into the polyethersulfone matrix membrane show better hydrophilicity and pure water flux.[170] Fouling resistance is exhibited due to the lower surface roughness and synergistic photocatalytic activity. The existence of nanotubes reduces the electron/hole recombination and improves photon efficiency. A multifunctional membrane designed by incorporating gold nanoparticles (AuNPs)/ exfoliated graphite nanoplatelets into a polysulfone membrane shows enhanced compaction resistance and permeability as well as superior catalytic property on the reduction of 4-nitrophenol to 4-aminophenol by $NaBH_4$, where AuNPs serve as the catalyst. The structure and catalytic activity of such membranes can be controlled separately by changing the relative contents of the corresponding components in the nanofiller hierarchy.

Nanocomposite membranes containing silica (SiO_2)/graphene oxide (GO) hybrid material[171, 172] exhibit better permeability, protein rejection and fouling resistance than SiO_2/polysulfone and GO/polysulfone membranes. The synergistic effect of SiO_2/GO is due to its high hydrophilicity as well as the special sandwiched structure that facilitates its dispersion in the polysulfone matrix.

Preparation of hybrid nafion membranes has been reported[173] using various fillers, such as titania nanoparticles, graphene oxide and organo-modified graphene oxide for water purification applications. The photocatalytic properties of the hybrid membranes have been evaluated using azo-dye methyl orange (MO) in aqueous solutions.

A partially reduced graphene oxide (rGO)/TiO_2 nanocomposite with five different molar ratios (rGO/TiO_2: 3/97, 30/70, 50/50, 70/30 and 90/10) has been synthesized[174] and characterized using X-ray diffraction and scanning electron microscopy (SEM) techniques. The polyvinylidene fluoride (PVDF) mixed-matrix membranes containing 0.05 wt% of the rGO/TiO_2 nanocomposite prepared by phase inversion show enhanced hydrophilicity, and better pure water flux and flux recovery ratio than the bare PVDF. A comparison of pure water flux of TiO_2- and rGO/TiO_2-containing membranes show that dispersion of inorganic TiO_2 nanoparticles on the GO surface decreases the aggregation of the nanoparticles and improves the characteristics of the mixed-matrix membrane. The blended PVDF membrane containing 0.05 wt% rGO/TiO_2 nanocomposite with GO to TiO_2 ratio of 70:30 shows higher permeability and antifouling characteristics. The hydrophobic property of PVDF membranes makes them prone to fouling; hydrophilic additives improve their hydrophilicity and antifouling performance. Silver nanoparticles prepared using PAMAM dendrimers as templates[175] to form silver-PAMAM dendrimer nanocomposites (Ag-DENPs) have good hydrophilicity and antibacterial performance. Ag-DENPs increase the surface roughness and decrease the pore size of PVDF membranes. Ag-DENPs provide significantly improved hydrophilicity of the PVDF membrane surface. Membrane permeation and antibacterial tests carried out to characterize the antifouling performance of PVDF membranes show that the flux recovery ratio increases by about 40% in the presence of Ag-DENPs on the PVDF membrane surface. The anti-organic fouling performance of PVDF membranes has improved.

4.3 THIN-FILM NANOCOMPOSITE (TFN)

A thin-film composite (TFC) membrane consists of a very thin barrier layer, normally made of polyamide, over a porous supporting layer. It is an interfacially synthesized membrane used for the removal of hardness, heavy metals and organic micro-pollutants such as pesticides, disinfection by-products (DBPs), etc. The development of forward osmosis (FO)/pressure-retarded osmosis (PRO) processes has boosted development of TFC membranes, as there is significant potential for further energy saving or even energy generation during desalination.[176–178]

Efforts have been made to select or modify the supporting layer to enhance adhesion between the barrier layer and the supporting layer, and to optimize the barrier layer by varying the interfacial polymerization (IP) conditions, i.e.,

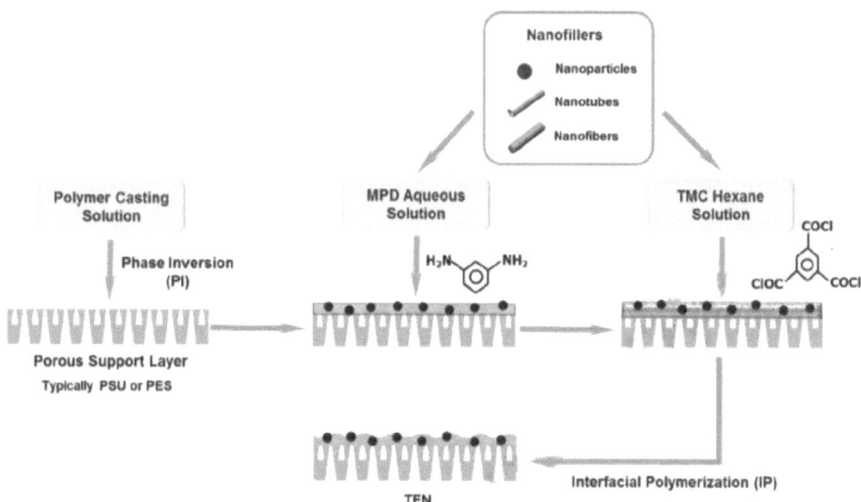

FIGURE 4.3 Preparation of thin-film nanocomposite (TFN) membrane by interfacial polymerization.

changing monomers, applying physical coating or chemical modification. With the advent of preparation technologies for nanocomposite materials, the concept of dispersing nanomaterials into the ultra-thin barrier to improve membrane performance for water purification has been proposed.[179] Nanomaterials used for making the conventional nanocomposite membrane have also been explored to prepare thin-film nanocomposite (TFN) membranes, including zeolites, carbon nanotubes, silica, silver and titania. The common method is the in-situ interfacial polymerization process between aqueous m-phenylenediamine (MPD) and trimesoyl chloride (TMC) organic solution (Figure 4.3). The nanofillers can be dispersed in either the aqueous or the organic phase.

4.3.1 Chlorine-Resistant Properties

The amide linkage in a polyamide thin-film composite (PA-TFC) membrane is susceptible to attack by chlorine, leading to degradation which adversely affects the life of the membrane. Therefore, care is taken to prevent the membrane from being exposed to strong oxidizing agents such as chlorine.[180] Solutions proposed include coating the polyamide surface with chlorine-resistant materials and introducing specific functional groups to the amide structure.[181]

Introducing nanomaterials into the polyamide (PA) structure provides a new dimension to chlorine-resistant membranes. When multi-wall nanotubes (MWNTs) are incorporated into the PA thin-film layer, the membrane shows improved chlorine resistance.[182] Interactions between the carboxylic group of modified MWNTs and the amide bond make the membrane more stable against chlorine. Similarly, the MWNT-embedded TFN membranes have a better chlorine resistance because the amide linkage is protected by electron-rich MWNTs.

Because amide bonds are the main target of chlorine attacks, the introduction of additional amide bonds or amino groups to the membrane enhances the protection of the PA cross-linking structure. Aminated zeolite[183] or hyper-branched polyamide modified silica nanoparticles[184] introduced into polyamide (PA) thin-film layers show enhanced chlorine resistance due to the intermolecular hydrogen bonding between aminated NPs and the PA structure, which mitigate the replacement of hydrogen by chlorine, and the additional amide bonds or amino groups introduced by aminated NPs that render the membrane more resistant to chlorine attack.

4.3.2 Thermal Stability

Thermal stability of the thin-film nanocomposite membrane improves with the incorporation of silica nanoparticles into the polyamide thin-film layer,[185, 186] due to the stronger electrostatic interaction between silica and polyamide in the modified polymer network structure. The incorporation of zeolite (NaX)[187] inside the polyamide structure also leads to a thin-film nanocomposite membrane with higher thermal stability and enhanced filtration performance. However, no significant change in the thermal decomposition temperature is noted, suggesting that there is no strong interaction between the nanofillers and polymer backbone chains. To improve the interaction between nanofiller and polyamide structure for further improvement of thermal stability, titania may be aminosilanized prior to the interfacial polymerization process,[188] resulting in a membrane that has better thermal stability, as well as improved permeability and selectivity.

4.3.3 Antifouling Properties

The introduction of hydrophilic nanofillers into the polyamide structure increases the surface hydrophilicity and helps to mitigate surface fouling. After incorporating hydrophilized ordered mesoporous carbons (H-OMCs) into polyamide (PA) thin-film layer, membranes show an enhanced surface hydrophilicity. It is also noted that the surface of the thin-film nanocomposite (TFN) membrane exhibits improved hydrophilicity after incorporation in situ of hydrophilic surface-modifying macromolecules.[189] Their 200 h fouling tests with sodium humate, silica particles and chloroform show that the TFN membranes have a much lower flux reduction than conventional thin-film composite (TFC) membranes. Decreased flux reduction is also observed when using a TFN membrane prepared with multi-wall nanotubes (MWNTs) to treat feed solution containing foulants.

4.3.4 Antibacterial Properties

Incorporation of nano-size antimicrobial and biocidal materials introduces anti-biofouling characteristics in nanocomposite membranes. With the addition of silver nanoparticles (AgNPs) to the polyamide structure during the interfacial polymerization process, the nanocomposite membrane can achieve appreciable anti-biofouling capability.[190] Incorporation of silver salts (such as silver citrate hydrate, silver lactate and silver nitrate) into the polyamide structure introduces antimicrobial capability into the membrane. Membranes prepared by dispersing silver nanoparticles in aqueous solution display quite good antibacterial properties.

Polyamide (PA) has a dense structure and good ion-rejection properties. Hence, silver nanoparticles (released as silver ions) embedded in the PA matrix are unlikely to be delivered to the interface between membrane and feed stream containing bacteria to act as the biocidal material. This kind of membrane may show a relatively slow leaching rate. It may not guarantee long-term antimicrobial activity. An attempt has been made to prepare an anti-biofouling membrane by developing silver nanoparticles immobilized on a thin-film composite polyamide membrane.[191] Development of a nano-silver and multi-walled carbon nanotubes thin-film nanocomposite membrane for enhanced water treatment has been pursued.[192] Work has been carried out to study the effect of mobile cations on a zeolite polyamide thin-film nanocomposite membrane.[193] Due to the significant biocidal activity of nanocrystals in silver form, it is anticipated that antimicrobial properties of the thin-film nanocomposite (TFN) membrane would improve. The TFN membrane exhibits higher water permeability and salt rejection but it does not show the anticipated strong biocidal activity.

4.3.5 Permeability and Selectivity

In order to overcome the trade-off between permeability and selectivity normally observed in polymeric membranes, it is desirable that solute rejection in the nanocomposite membrane case should remain nearly the same while water permeability is improved. In the thin-film composite (TFC) membrane, the polyamide thin-film layer controls aspects of membrane performance such as permeability, selectivity and fouling resistance. The incorporation of nanomaterials into the polyamide layer improves the physico-chemical properties of the membrane such as hydrophilicity, charge density, porosity and cross-linking. It provides a kind of water channel with the potential to overcome the permeability/selectivity trade-off relationship.

According to solution-diffusion theory, an increase in the hydrophilicity of the membrane facilitates water solubilization and diffusion through the membrane, improving water permeability.[194, 195] Hydrophilic nanofillers result in a thin-film nanocomposite membrane with reduced contact angle, indicating enhanced surface hydrophilicity. For example, the contact angle of a zeolite-polyamide thin-film nanocomposite membrane decreases from around 70° to 40° with increasing zeolite loading from 0% to 0.4% in the organic phase. The contact angle for the oxidized multi-wall-nanotubes polyamide thin-film nanocomposite membrane decreases from around 70° to 25° with increasing multi-wall-nanotube loading from 0% to 0.2% in the aqueous phase. For mesoporous silica polyamide thin-film nanocomposite membranes, the contact angle decreases from around 57° to 28° with increasing silica loading from 0% to 0.1% in the organic phase.[196] These studies demonstrate enhanced water permeability with increasing nanofiller loading.

The decrease in the contact angle in the presence of embedded nanoparticles is caused by hydration and hydrophilicity. As well as hydrophilicity, the cross-linking condition and thickness of the thin-film layer are also important parameters controlling water permeability and salt rejection.[197] In general, a lower degree of cross-linking and reduced thickness of the thin-film layer result in higher water permeability.

Embedding nanofillers in the polyamide matrix can reduce the cross-linking in the thin-film layer by disturbing the reaction between amine groups and acyl chloride groups or even forming nanovoids around the interfaces between the nanofiller and polyamide matrix.

Thin-film nanocomposite membranes are less cross-linked than thin-film composite membranes; however, the extent of cross-linking does not show a strong correlation with water or salt permeability. It suggests that defects or the molecular sieving effect might have played a major role in the separation performance of the membrane. The effect of nano-crystal size on thin-film nanocomposite membrane performance is quite appreciable. The nanogaps between the zeolite crystals and the polymer matrix can provide a low resistance pathway for solvent, enhancing permeability.[198] The polyamide matrix appears to be the main contributor to the solute rejection.

In addition to causing changes to the polymer cross-linking, the incorporation of nanofillers provides additional channels for the transport of water and not for the solutes. Incorporation of zeolite nanoparticles in a thin-film nanocomposite membrane results in higher water flux without compromising over salt rejection. The hydrophilic nanoparticles provide preferential flow paths for water molecules. Compared to the non-porous silica nanoparticles, mesoporous silica nanoparticles containing highly ordered hexagonal pores have a significantly higher impact on water permeability.[199] With non-porous silica nanoparticles, a reduction in polymer cross-linking can still occur, but there will be no water permeation through the internal structures of nanoparticles; therefore, while the observed water flux in the membrane containing non-porous silica is higher than in the conventional thin-film composite, it is less than that in the membrane containing mesoporous silica. It indicates that the internal pores of nanofiller contribute significantly to enhanced water permeability. The high solute rejection can be maintained by a combination of steric and Donnan exclusion.[200] So the goal of breaking the trade-off relationship mainly relies on one or more factors, such as a moderate reduction in cross-linking, an enhanced membrane surface charge density, appropriate nanovoids and additional water channels. Zeolite can provide molecular sieving channels and enhanced charge density. Mesoporous silica can provide large water channels combined with enhanced charge density, while aquaporin can provide exclusive water channels.

To fully utilize the favorable properties of nanomaterials, nanofillers of appropriate size, internal structure and surface properties are introduced. Suitable interfacial interactions between them and polymer matrices are ensured. Typical thickness of the thin-film layer is a few hundred nanometers. Small nanoparticles (up to 100 nm in size) in the thin-film nanocomposite membrane are effective in enhancing its permeability. To improve the dispersion of nanofillers and the interaction between nanofillers and the polymer matrix, the surface of nanofillers is commonly modified prior to the embedding process. For example, carbon nanotubes are treated with concentrated acid to generate oxygen-containing functional groups so that adequate dispersion can be achieved for the preparation of the membrane. Good interactions between carbon nanotubes and polymer matrices also need to be provided.

4.3.6 THIN-FILM COMPOSITE WITH NANOCOMPOSITE SUBSTRATE

In a thin-film composite membrane with nanocomposite substrate membrane, silica or zeolite nanoparticles are embedded into the polysulfone substrate, which is then used in the interfacial polymerization process to prepare a thin-film composite membrane.[201] This type of membrane shows higher initial permeability and lower flux decline during compaction than the original thin-film composite membrane. The nanomaterial provides the necessary mechanical support to mitigate the collapse of the porous structure and to resist thickness reduction due to compaction. Membranes with a nanocomposite substrate undergo less physical compaction and play an important role in maintaining high water permeability (Figure 4.4).[202]

This decreases internal concentration polarization (ICP) inside the porous support layer, reducing the probability of a potential adverse impact on the forward osmosis and pressure-retarded osmosis processes because it can significantly reduce the available osmotic driving force and hence lower the water flux. The nanocomposite substrate has enhanced hydrophilicity. Incorporation of zeolite into the polysulfone substrate reduces internal concentration polarization and provides higher water flux due to improved porosity, better hydrophilicity and additional water pathways through porous nanoparticles. It offers the opportunity to use porous nanoparticles and nanocomposite substrate to control the internal concentration polarization in forward osmosis operation. Subsequent studies have been conducted using multi-walled carbon nanotubes[203] and TiO_2[204–206] to mitigate the problem of

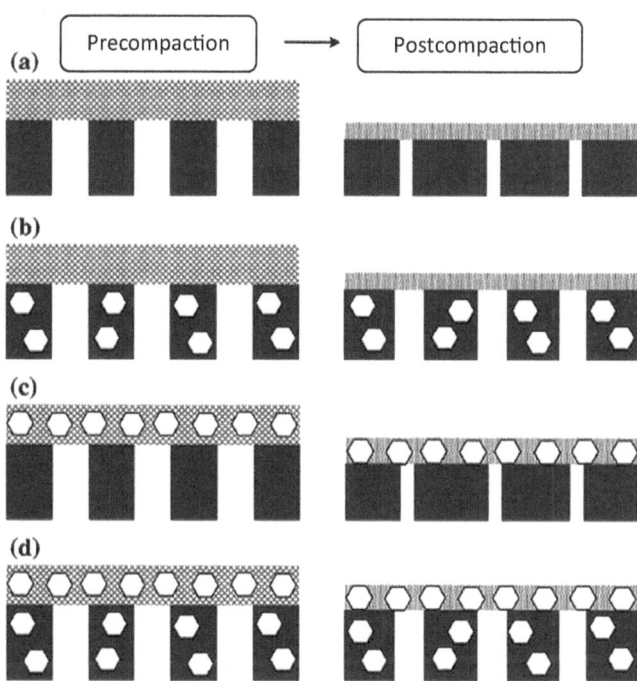

FIGURE 4.4 Incorporation of nanoparticles in TFC substrate leading to decreased compaction.

internal concentration polarization in the nanocomposite substrate for forward osmosis process. The resulting membranes have shown enhancement in performance of forward osmosis and a reduction in internal concentration polarization. The same substrate also shows better tensile strength, which is a desirable property for many applications.

4.4 BIOINSPIRED MEMBRANES

In this section, the work carried out in the field of bioinspired membranes is discussed, with particular reference to vertically aligned carbon nanotube membranes, graphene and aquaporin membranes.

4.4.1 Carbon Nanotube Membrane

Because of their outstanding properties, carbon nanotubes (CNTs) are regarded as potential candidates for diverse nano-technological applications, such as fillers in polymer matrices, sensors and others. It provides higher flux for water passing through carbon nanotube channels, better selectivity by side-wall and tip functionalization, and antifouling characteristics for both hydrophilic and hydrophobic fouling as well as biofouling by selective chemical modification of carbon nanotube architecture. Research in the field of carbon nanotubes[207] took a quantum leap forward following the synthesis of fullerenes. CNTs have high potential for the removal of lead, cadmium, fluoride and arsenic from contaminated water.[208–211] The alignment and functionalization of carbon nanotubes are important in deciding the sorption capacity of carbon nanotubes. The macro-architecture of aligned carbon nanotubes has also been reported for removal of petroleum products and microorganisms.[212] However, scaling up is still a challenging issue in carbon nanotube-embedded membranes.[213]

4.4.1.1 Functionalization of Carbon Nanotubes

Functionalization of carbon nanotubes has the potential to open up wider avenues for utilizing the benefits of carbon nanotubes and gain better separation characteristics as well as high throughput. In addition, attaching organic moieties facilitates better anchoring of nanotubes in host materials and yields better reinforcement of composites. The main approaches to the modification of carbon nanotubes can be grouped into two categories: (a) the covalent attachment of chemical groups through reactions onto the π-conjugated skeleton of the carbon nanotube; and (b) the endohedral filling of their inner empty cavity.

An important breakthrough in nanotube chemistry was the oxidation of a carbon nanotube in concentrated nitric acid.[214] Such a drastic condition helps open the carbon nanotube tips as well as oxidative etching along the sidewalls. This enables the decoration of walls with various oxygen-containing groups (mainly the carboxyl group). The incorporation of the carboxyl group exposes useful sites in carbon nanotubes for further modification as per the requirement. In addition, the formation of anhydride can take place at the tube ends through which the rings of carbon nanotubes are accessible.[215] The introduction of the carboxyl group reduces the van der Waal forces existing between the individual carbon nanotubes; hence the carbon nanotubes can be made water soluble by addition/substitution of new moieties because carbon nanotubes as grown are not soluble in any solvent.

Addition reactions like fluorination,[216–218] hydrogeneation[219] and cycloaddition[220–224] help in direct coupling of functional groups onto the π-conjugated carbon framework. The fluorine atoms of fluorinated CNTs can be replaced through nucleophilic substitution reaction, and thus, functional groups of alcohols, amines, etc. can be successfully incorporated on the CNT sidewall.[225, 226]

4.4.1.2 Wetting and Filling of CNT Cavity

The wetting properties of CNT determine which liquid will fill the tube by capillary action and cover the inner surface. The Young–Laplace equation relates the pressure difference (ΔP) across the liquid–vapor interface in a capillary to the surface tension of the liquid and the contact angle between solid and liquid:

$$\Delta P = 2\gamma\, r^{-1} \cos\theta \tag{4.1}$$

where r is the radius of curvature of the meniscus. The contact angle θ is an indicator of the strength of the interaction between the liquid and the solid interface relative to the cohesive forces in the liquid. If θ is smaller than 90°, the contact between the liquid and the surface is said to be wetting and ΔP is positive. Therefore, liquid will be pulled into the capillary spontaneously as there is an energy gain in the wetting process. If θ is greater than 90°, the contact angle is said to be nonwetting and ΔP will be negative. Therefore, when $\theta > 90°$, the only way to introduce liquid into a capillary is to apply pressure larger than (ΔP).

The wetting properties of the carbon nanotubes measured experimentally[227] imply that there is a cut-off point in the surface tension of the liquid above which wetting no longer occurs. The capillary action of nanotubes observed[228–230] for oxides indicates that the surface tension values of compounds are below the cut-off point determined.

If the surface tension of the liquid or molten salt is sufficiently low, wetting occurs and the carbon nanotubes can be filled by capillarity. However, to fill carbon nanotubes with higher surface-tension materials (like metals or metal oxides), one has to force the molten materials into open carbon nanotubes by applying large pressure followed by cooling of the sample before pressure is dropped, so that the material is trapped inside. The convenient way is to use a low surface-tension solvent carrier. In such a case, the compound can be dissolved in nitric acid which would open up the tips of carbon nanotubes, and the material is carried into the empty cavity of carbon nanotubes by capillarity.

4.4.1.3 Mass Transport Through CNT Channels

High aspect ratio, molecularly smooth hydrophobic graphitic walls and nanoscale inner diameters of carbon nanotubes give ultra-efficient transport of water and gas through these ultra-narrow molecular pipes. Water and gas molecules move through nanotube pores much faster. The water transport mechanism is similar to those of biological ion channels.

Because of the narrow diameter of carbon nanotubes, thermodynamic and transport properties of confined water differ substantially from those observed in the bulk.[231]

When simulated within narrow molecular sieves, diffusion is dominated by concerted events in which multiple molecules move simultaneously.[232] These concerted events are due to strong mismatches in (i) the distance between binding sites along the pore axis, and (ii) distance between adsorbed molecules, which minimizes adsorbate–adsorbate interactions.[233] On the other hand, the inherent smoothness of the interior of carbon nanotubes generates exceptionally high diffusion coefficients for water.[234–236] Water molecules occasionally move along the single-wall nanotube (0.8 nm diameter and 1.35 nm long) axis via bursts of hydrogen-bonded clusters of molecules.[237]

The average flow velocity (υ) of an incompressible, creeping liquid (i.e., Reynolds number well below one) inside a tube with a uniform cross-sectional area is given by Darcy's Law:

$$\upsilon = \gamma\left(\Delta P/L\right) \tag{4.2}$$

where ΔP is the pressure difference across the tube, L is its length, and γ is the hydraulic conductivity. Although Darcy's Law is an empirical expression, the hydraulic conductivity of a Newtonian liquid in a circular tube subject to the no-slip boundary condition, $\gamma_{no\text{-}slip}$, can be found directly from the no-slip Poiseuille relation.[238]

Liquid slip at the solid-liquid boundary, confinement-induced reductions in the liquid viscosity and subcontinuum changes to the liquid structure can cause the actual hydraulic conductivity (γ_{actual}) to exceed the calculated hydraulic conductivity from the Poiseuille relation.[239] This increase in γ leads to the definition of a flow-enhancement factor, ε,

$$\varepsilon = \gamma_{actual}/\gamma_{no-slip}. \tag{4.3}$$

For carbon nanotubes with diameters larger than 1.6 nm, the variation in hydraulic conductivity that occurs with carbon nanotube diameter can be understood in terms of slip at the water–carbon nanotube boundary and diameter-related changes to the water viscosity. In carbon nanotubes with smaller diameters, however, water molecules have been shown to assemble into diameter-dependent one-dimensional structures for which neither the slip length nor the effective viscosity is well defined. This confinement-induced change to the liquid structure necessitates a subcontinuum description of the liquid.[240]

4.4.1.4 Antimicrobial Properties of Carbon Nanotubes

Studies have been carried out on single-wall nanotubes with antimicrobial activity towards gram-positive and gram-negative bacteria and the damages inflicted are attributed to either a physical interaction with or oxidative stress on cell membrane integrity.[241, 242] Carbon nanotubes may therefore be useful for inhibiting microbial attachment and biofouling on surfaces. However, the degree of aggregation[243] and the bioavailability of the nanotubes[244] need to be considered to exploit the antimicrobial properties effectively.

4.4.1.5 Preparation of Carbon Nanotube-Based Membrane

In the ex-situ alignment method, the carbon nanotubes are aligned in advance using carbon vapor deposition (CVD). They are then compounded with the polymer matrix by either in-situ polymerization of some monomers or by spin coating/dip coating of polymer solution onto the aligned carbon nanotube matrix. There are challenges associated with each step of membrane making, from growth of carbon nanotubes to membrane performance evaluation and scale-up. There are four main approaches to the synthesis of carbon nanotube-based membranes[245] (Figure 4.5).

a) The template synthesis approach, in which carbonaceous material is deposited inside pre-existing ordered porous membranes such as anodized alumina. It is also known as template synthesized carbon nanotube membranes.[246] A scanning electron micrograph (SEM) of the nanotubes after dissolution of the template is shown in Figure 4.5a.

b) Membranes based on the interstice between nanotubes in a vertical array of carbon nanotubes, referred to as the dense-array outer-wall carbon nanotube membrane in which the fluid is transported through the interstice between the nanotubes, although some transport can occur through open-ended tubes. The SEM image in Figure 4.5b demonstrates the dense array of carbon nanotubes.

c) Encapsulation of as-grown vertically aligned carbon nanotubes by a space-filling inert polymer or ceramic matrix followed by opening up the carbon nanotube tips using plasma chemistry, or the open-ended carbon nanotube

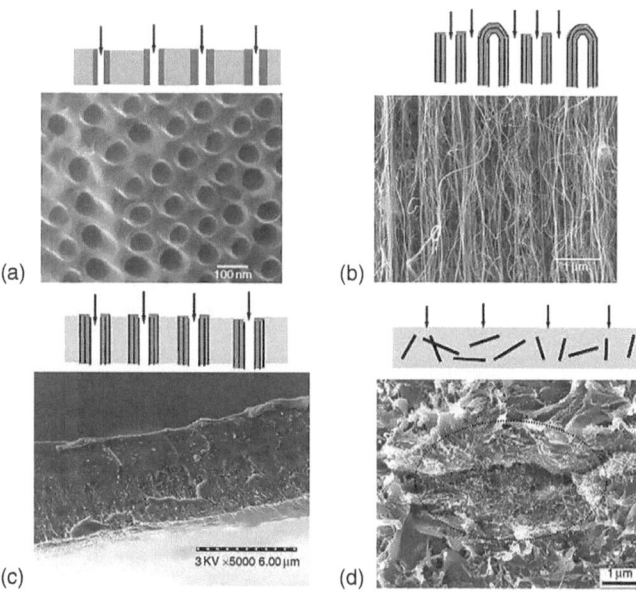

FIGURE 4.5 Different approaches to carbon nanotube membrane synthesis: (a) template synthesis approach, (b) dense-array outer-wall CNT membrane approach, (c) encapsulation of as-grown vertically aligned carbon nanotubes by a space-filling inert polymer or ceramic matrix followed by opening up the carbon nanotube tips, (d) membranes composed of nanotubes as fillers in a polymer matrix.

membrane[247, 248] in which fluid transport occurs through the inner core of the carbon nanotube. The SEM image shows the cross-section of the membrane with aligned carbon nanotube in an impervious polymer matrix (Figure 4.5c).

d) Membranes composed of nanotubes as fillers in a polymer matrix, also known as mixed-matrix membranes. An SEM image of the composite membrane structure is shown in Figure 4.5d.

In the case of the carbon nanotube membrane, the pressure-driven flux is several times higher than conventional Newtonian flow due to atomically flat graphite planes inducing nearly ideal slip conditions.[249]

Membranes with carbon nanotube tips that were functionalized with biotin show[250] a reduction in $Ru(NH_3)_6$ by a factor of 15 when bound with streptavidin, demonstrating the potential for molecular transport through carbon nanotube cores for applications in chemical separation. The embedded carbon nanotube in the host membrane matrix works on the size-exclusion principle. However, the functionalization of the carbon nanotube tip can introduce the required physico-chemical characteristics into the membrane surface, which may lead to selective removal of contaminants based on physico-chemical interaction of species with the functional group present over the carbon nanotube tip.

Antimicrobial activity of carbon nanotubes requires direct contact between carbon nanotubes and target microorganisms.[251] Suspension of non-functionalized carbon nanotubes in water is difficult and does not provide enough contact for disinfection. Accordingly, the antibacterial activity of carbon nanotubes can be utilized by coating them on a reactor surface in contact with the pathogen-laden water. Complete retention and effective inactivation of E-coli can be achieved using a PVDF microporous membrane coated with a thin layer of single-wall nanotubes.[252] Although the rate of bacterial inactivation by carbon nanotube is relatively low compared to conventional disinfectants, it is sufficient to prevent biofilm formation and the subsequent biofouling of the membrane surface. This increases the lifespan of the membrane without significant decline in throughput because of biofouling. In addition, the selective functionalization of the carbon nanotube tip with hydrophilic or hydrophobic groups can help minimize fouling depending upon the feed water conditions.

Though carbon nanotubes have great potential to offer an ideal nanocomposite membrane, there are several issues.[253, 254] It is challenging to grow 12–13 order of magnitude (i.e. 10^{12} to 10^{13}) of carbon nanotubes per square centimeter area. Chemical vapor deposition methodology offers good results. It is difficult to achieve a carbon nanotube yield beyond 90% in a particular batch of synthesis. The purification steps to remove the deposits and to make the carbon nanotube wall defect free are quite challenging. During preparation of the nanocomposite membrane, carbon nanotubes must be well-dispersed and well-aligned. Functionalization of carbon nanotubes with the desired functional groups may be required to achieve better dispersion, which is quite challenging because the carbon nanotube is not soluble. Opening of the carbon nanotube tips is performed by either acid treatment or plasma-based oxidation. This is a critical step and is not simple to achieve. The tip-opening step may result in thinning of the carbon nanotube wall and disruption of the tube integrity, and may subsequently lead to failure of membrane channels. Scale-up of carbon nanotube growth, carbon nanotube alignment, nanocomposite formation, carbon nanotube tip or side-wall functionalization is quite complex, involving material as well as process challenges.

4.4.2 GRAPHENE MEMBRANES

The synthesis of nanoporous graphene (NPG) and graphene oxide (GO) membranes opens the door for next-generation membranes as a promising alternative to thin-film composite polyamide membranes for water purification. Graphene has the potential for mechanically robust, ultra-thin, high-flux, high-selectivity and fouling-resistant membranes.[255, 256] Graphene-based materials have attracted interest for use in membranes for desalination and water purification due to their unique properties, distinctive structural characteristics, high mechanical strength and low thickness.[257–259] Advances in molecular simulation of the graphene family have potential for developing novel membrane desalination technologies. Graphene's unique electronic properties, high tensile strength and impermeability to small molecules have been established[260–264] and utilized to make thin membranes with size-tunable pores (for molecular sieving) allowing for high flux. Graphene nanosheets display favorable chemical and physical properties in the desalination process. Despite its negligible thickness, membranes made of graphene exhibit adequate mechanical strength and ability to function under higher pressures, making them superior to conventional polymeric reverse osmosis membranes.

Simulation studies have identified nanoporous graphene (NPG) structures as among the most promising membrane materials that can provide high water flow rates and high salt rejection as a function of nanopore morphology.[265] On the other hand, these hypotheses are based on a single layer of graphene sheet which is difficult to assemble in practice.[266] The transport of water through these nanoporous membranes can reach up to 66 L/cm^2/day/MPa with greater than 99% salt rejection. In contrast, water transport through a conventional reverse osmosis membrane reaches approximately 0.01–0.05 L/cm^2/day/MPa with similar salt rejection. These values reveal the potential for the utilization of functionalized NPG as a high-permeability desalination membrane (Figure 4.6a and b).

Figure 4.6a shows high-pressure molecular and ionic sieving across a one-atom-thick graphene sheet. Chemical functionalization of the pores with hydrogen increases water selectivity, whereas functionalization with hydroxyl groups improves the flux.[267] Performance of functionalized nanoporous graphene and membranes used in existing desalination technologies is shown in Figure 4.6b. It is noted that the performance of functionalized nanoporous graphene is superior both from water permeability and salt rejection points of view to conventional membranes used in desalination plants.

With current advances, large areas of graphene have been successfully grown on plain copper foils at atmospheric pressure.[268] The feasibility of mass-producing graphene has widened the application spectrum. Graphene monolayers with superior strength can allow water to flow through them while hindering the passage of other unwanted species.[269, 270] Computational studies have predicted that large slip length

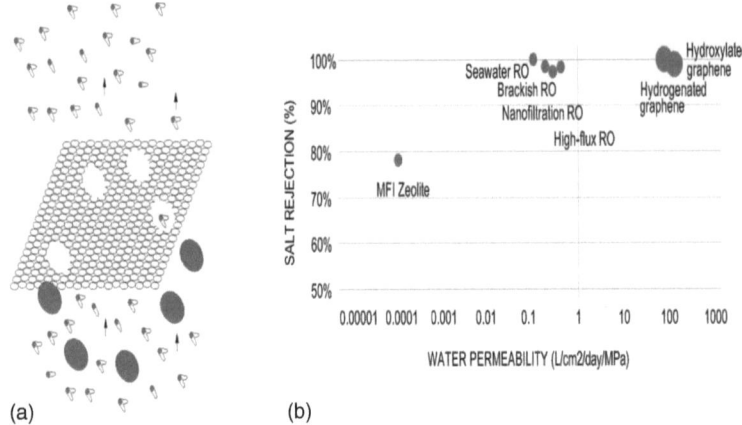

(a)   (b)

FIGURE 4.6 (a) High-pressure molecular and ionic sieving across a one-atom-thick graphene sheet. (b) Performance of functionalized nanoporous graphene and membranes used in existing desalination technologies.

of water molecules allow negligible friction with graphene surfaces, which results in an ultra-fast water flow in the graphene nanochannel,[271–274] providing much higher membrane flux than in commercially available reverse osmosis and nanofiltration membranes.[275]

4.4.3 AQUAPORIN MEMBRANES

Biomimetics and bioinspiration have great potential in exploring membrane materials and membrane processes. Due to their unique combination of high water permeability and high solute rejection, aquaporin proteins have attracted lot of interest as functional building blocks of biomimetic membranes.[276–278]

Aquaporins are the protein channels that control water flux across biological membranes.[279] Aquaporin is found widely in human tissues for rapid passive transport of water across cell membranes. Such transport channels exist in the cells of species in all three domains of life. Water movement in aquaporins is mediated by selective rapid diffusion caused by osmotic gradients.[280] The hourglass shape of aquaporin-1 (AQP1), with selective extracellular and intracellular vestibules at each end, allows water molecules to pass rapidly in single file, while excluding proteins.[281] A constant number of molecules are assumed to occupy the aquaporin channel at all times and the water molecules are assumed to move together in discrete translocations, or hops.[282]

The highly selective water permeability of aquaporin channels is an interesting property when considering water treatment membranes. Transport of water maintaining selectivity through biological lipid bilayers, containing aquaporins, surpasses the performance of conventional reverse osmosis membranes. Single aquaporins transport water molecules at 2–8 billion molecules per second. Kaufman et al. predict that a membrane with 75% coverage of aquaporins could have a hydraulic permeability

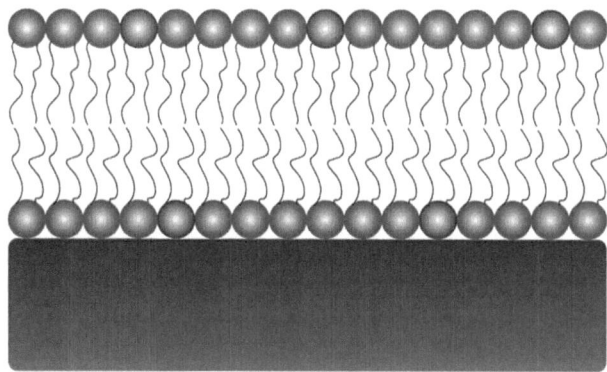

FIGURE 4.7 Conceptual cross-sectional image of a semipermeable lipid bilayer membrane cast atop a nanofiltration-type support membrane.

in the range of 2.5×10^{-11} m $Pa^{-1}s^{-1}$, an order of magnitude higher than a conventional seawater reverse osmosis membrane.[283] Transport across biological membranes is driven by salt concentration gradient or osmotic pressure gradient. A conceptual cross-sectional image of a semipermeable lipid bilayer membrane cast on a nanofiltration-type membrane is shown in Figure 4.7.

Water transport through the aquaporin-embedded vesicular membrane has been explored in the forward osmosis (FO) process.[284] Mathematical simulation correlates vesicle size, vesicle permeability and the interior solute concentration with membrane flux in the forward osmosis process. It indicates that the water flux of the membrane can be determined by vesicle size and permeability in pressure-retarded mode. The interior solute concentration of vesicles has an impact on the hydrostatic pressure of the vesicles in forward osmosis mode.

Planar biomimetic membranes consisting of aquaporin Z (AqpZ) are made[285] on cellulose acetate membrane substrate and functionalized with methacrylate end groups. A thin-film composite (TFC) aquaporin-based biomimetic membrane (ABM) is prepared[286] by the interfacial polymerization method, in which AquaporinZ-containing proteoliposomes is added to the m-phenylene-diamine aqueous solution. Control membranes, either without aquaporins or with inactive (mutant) aquaporins, can also be similarly made. The separation performance of the membranes is evaluated by cross-flow reverse osmosis (RO) tests. The active ABM gives higher water permeability (4 L/m²·h·bar) with comparable NaCl rejection (97%) at 5 bar pressure. Its permeability is about 40% more than a conventional brackish water reverse osmosis membrane and an order of magnitude higher than a sea water reverse osmosis membrane. This implies great potential of the TFC-ABM for desalination applications. In another study,[287] proteoliposome containing aquaporin is fully encapsulated into the thin-film layer through cross-linking of poly(ethyleneimine) (PEI). This novel TFN membrane shows significantly enhanced water permeability and typical salt rejection with respect to $MgCl_2$.

4.5 CHALLENGES AND OPPORTUNITIES

Development of polymer-matrix nanocomposite membranes for water treatment is on the increase. The incorporation of nanomaterial can transform membranes with unique properties. It has the potential to induce new characteristics based on their synergistic effects and opens up a new dimension to design of advanced polymeric membranes with better performance and antifouling properties. The potential applications of nanocomposite membranes are huge, covering the whole filtration spectrum including microfiltration, ultrafiltration, nanofiltration, reverse osmosis, forward osmosis, etc. Several challenges still need to be addressed to optimize and commercialize nanocomposite membranes for industrial applications. Work on the effect of nanomaterials on membrane structure and performance is needed. There is also a need to examine the dispersion of nanomaterials as well as the compatibility of nanofillers with polymers to avoid nanomaterials leaching into the environment. Large-scale production issues and the techno-economics of applications also need to be evaluated.[288]

REFERENCES

1. Yin, J., Deng, B. 2015. Polymer-matrix nanocomposite membranes for water treatment. *J. Membr. Sci.* 479: 256–275.
2. Jia, M., Peinemann, K.V., Behling, R.D. 1991. Molecular sieving effect of the zeolite-filled silicone rubber membranes in gas permeation. *J. Membr. Sci.* 57: 289–296.
3. Duval, J.M., Folkers, B., Mulder, M.H.V., Desgrandchamps, G., Smolders, C.A. 1993. Adsorbent filled membranes for gas separation. Part 1. Improvement of the gas separation properties of polymeric membranes by incorporation of microporous adsorbents. *J. Membr. Sci.* 80: 189–198.
4. Wu, H., Tang, B., Wu, P. 2010. Novel ultrafiltration membranes prepared from a multi-walled carbon nanotubes/polymer composite. *J. Membr. Sci.* 362: 374–383.
5. Goh, P.S., Ng, B.C., Ismail, A.F., Aziz, M., Sanip, S.M. 2010. Surfactant dispersed multi-walled carbon nanotube/polyetherimide nanocomposite membrane. *Solid State Sci.* 12: 2155–2162.
6. Shawky, H.A., Chae, S.R., Lin, S., Wiesner, M.R. 2011. Synthesis and characterization of a carbon nanotube/polymer nanocomposite membrane for water treatment, *Desalination* 272: 46–50.
7. Mansourpanah, Y., Madaeni, S.S., Rahimpour, A., Adeli, M., Hashemi, M.Y., Moradian, M.R. 2011. Fabrication new PES-based mixed matrix nanocomposite membranes using polycaprolactone modified carbon nanotubes as the additive: Property changes and morphological studies. *Desalination* 277: 171–177.
8. Vatanpour, V., Madaeni, S.S., Moradian, R., Zinadini, S., Astinchap, B. 2011. Fabrication and characterization of novel antifouling nanofiltration membrane prepared from oxidized multiwalled carbon nanotube/polyethersulfone nanocomposite. *J. Membr. Sci.* 375: 284–294.
9. Majeed, S., Fierro, D., Buhr, K., Wind, J., Du, B., Boschetti-de-Fierro, A., Abetz, V. 2012. Multi-walled carbon nanotubes (MWCNTs) mixed polyacrylonitrile (PAN) ultrafiltration membranes. *J. Membr. Sci.* 403–404: 101–109.
10. Zhao, X., Ma, J., Wang, Z., Wen, G., Jiang, J., Shi, F., Sheng, L. 2012. Hyperbranched-polymer functionalized multi-walled carbon nanotubes for poly (vinylidene fluoride) membranes: From dispersion to blended fouling-control membrane. *Desalination* 303: 29–38.

11. De Lannoy, C.F., Jassby, D., Davis, D.D., Wiesner, M.R. 2012. A highly electrically conductive polymer multiwalled carbon nanotube nanocomposite membrane. *J. Membr. Sci.* 479(415–416): 718–724.
12. Daraei, P., Madaeni, S.S., Ghaemi, N., Ahmadi Monfared, H., Khadivi, M.A. 2013. Fabrication of PES nanofiltration membrane by simultaneous use of multi-walled carbon nanotube and surface graft polymerization method: Comparison of MWCNT and PAA modified MWCNT. *Sep. Purif. Technol.* 104: 32–44.
13. Daraei, P., Madaeni, S.S., Ghaemi, N., Khadivi, M.A., Astinchap, B., Moradian, R. 2013. Enhancing antifouling capability of PES membrane via mixing with various types of polymer modified multi-walled carbon nanotube. *J. Membr. Sci.* 444: 184–191.
14. De Lannoy, C.F., Soyer, E., Wiesner, M.R. 2013. Optimizing carbon nanotube-reinforced polysulfone ultrafiltration membranes through carboxylic acid functionalization. *J. Membr. Sci.* 447: 395–402.
15. Yin, J., Zhu, G., Deng, B. 2013. Multi-walled carbon nanotubes (MWNTs)/polysulfone (PSU) mixed matrix hollow fiber membranes for enhanced water treatment. *J. Membr. Sci.* 437: 237–248.
16. Phao, N., Nxumalo, E.N., Mamba, B.B., Mhlanga, S.D. 2013. A nitrogen-doped carbon nanotube enhanced polyethersulfone membrane system for water treatment, *Phy. Chem.* 66: 148–156.
17. Zirehpour, A., Rahimpour, A., Jahanshahi, M., Peyravi, M. 2014. Mixed matrix membrane application for olive oil wastewater treatment: Process optimization based on Taguchi design method. *J. Environ. Manage.* 132: 113–120.
18. Saranya, R., Arthanareeswaran, G., Dionysiou, D.D. 2014. Treatment of paper mill effluent using Polyethersulfone/functionalized multiwalled carbon nano-tubes based nanocomposite membranes. *Chem. Eng. J.* 236: 369–377.
19. Badawi, N.E., Ramadan, A.R., Esawi, A.M.K., Mohamed, E.M. 2014. Novel carbon nanotube–cellulose acetate nanocomposite membranes for water filtration applications. *Desalination* 344: 79–85.
20. Nezam El-Dein, L.A., El-Gendi, A., Ismail, N., Abed, K.A., Ahmed, A.I. 2014. Evaluation of cellulose acetate membrane with carbon nanotubes additives. *J. Ind. Eng. Chem.* 26: 259–264.
21. Kim, E-S., Liu, Y., Mohamed Gamal, E.D. 2013. An in-situ integrated system of carbon nanotubes nanocomposite membrane for oils and process-affected water treatment. *J. Membr. Sci.* 429: 418–427.
22. Mauter, M.S., Elimelech, M. 2008. Environmental applications of carbon-based nano-materials. *Environ. Sci. Technol.* 42: 5843–5859.
23. Wang, X., Chen, X., Yoon, K., Fang, D., Hsiao, B.S., Chu, B. 2005. High flux filtration medium based on nanofibrous substrate with hydrophilic nanocomposite coating. *Environ. Sci. Technol.* 39: 7684–7691.
24. Goh, K., Setiawan, L., Wei, L., Jiang, W., Wang, R., Chen, Y. 2013. Fabrication of novel functionalized multi-walled carbon nanotube immobilized hollow fiber membranes for enhanced performance in forward osmosis process. *J. Membr. Sci.* 446: 244–254
25. Choi, J.H., Jegal, J., Kim, W.N. 2006. Fabrication and characterization of multi-walled carbon nanotubes/ polymer blend membranes. *J. Membr. Sci.* 284: 406–415.
26. Tang, C., Zhang, Q., Wang, K., Fu, Q., Zhang, C. 2009. Water transport behavior of chitosan porous membranes containing multi-walled carbon nanotubes (MWNTs). *J. Membr. Sci.* 337: 240–247.
27. Majeed, S., Fierro, D., Buhr, K., Wind, J., Du, B., Boschetti-de-Fierro, A., Abetz, V. 2012. Multi-walled carbon nanotubes (MWCNTs) mixed polyacrylonitrile (PAN) ultrafiltration membranes. *J. Membr. Sci.* 403–404: 101–109
28. Ebert, K., Fritsch, D., Koll, J., Tjahjawiguna, C. 2004. Influence of inorganic fillers on the compaction behaviour of porous polymer based membranes. *J. Membr. Sci.* 233: 71–78.

29. Bae, T.H., Tak, T.M. 2005. Effect of TiO$_2$ nanoparticles on fouling mitigation of ultrafiltration membranes for activated sludge filtration. *J. Membr. Sci.* 249: 1–8.
30. Arsuaga, J.M., Sotto, A., Rosario, G.D., Martínez, A., Molina, S., S.B. Teli, J. de Abajo. 2013. Influence of the type, size, and distribution of metal oxide particles on the properties of nanocomposite ultrafiltration membranes. *J. Membr. Sci.* 428: 131–141.
31. Yang, Y., Wang, P., Zheng, Q. 2006. Preparation and properties of polysulfone/TiO$_2$ composite ultrafiltration membranes. *J. Polym. Sci. Part B: Polym. Phys.* 44: 879–887.
32. Cao, X., Ma, J., Shi, X., Ren, Z. 2006. Effect of TiO$_2$ nanoparticle size on the performance of PVDF membrane. *Appl. Surf. Sci.* 253: 2003–2010.
33. Xiao, Y., Yu, K., Wang, T.S., Chung Tan, J. 2006. Evolution of nano-particle distribution during the fabrication of mixed matrix TiO$_2$-polyimide hollow fiber membranes. *Chem. Eng. Sci.* 61: 6228–6233.
34. Li, J.B., Zhu, J.W., Zheng, M.S. 2007. Morphologies and properties of poly(phthalazinone ether sulfone ketone) matrix ultrafiltration membranes with entrapped TiO$_2$ nanoparticles. *J. Appl. Polym. Sci.* 103: 3623–3629.
35. Yang, Y., Zhang, H., Wang, P., Zheng, Q., Li, J. 2007. The influence of nano-sized TiO$_2$ fillers on the morphologies and properties of PSF UF membrane. *J. Membr. Sci.* 288: 231–238.
36. Fu, X., Matsuyama, H., Nagai, H. 2008. Structure control of asymmetric poly(vinyl butyral)-TiO$_2$ composite membrane prepared by nonsolvent induced phase separation. *J. Appl. Polym. Sci.* 108: 713–723.
37. Wu, G., Gan, S., Cui, L., Y. Xu. 2008. Preparation and characterization of PES/TiO$_2$ composite membranes. *Appl. Surf. Sci.* 254: 7080–7086.
38. Yu, L.Y., H.M. Shen, Xu, Z.L. 2009. PVDF-TiO$_2$ composite hollow fiber ultrafiltration membranes prepared by TiO$_2$ sol-gel method and blending method. *J. Appl. Poly. Sci.* 113: 1763–1772.
39. Li, J.F., Xu, Z.L., Yang, H., Yu, L.Y., Liu, M. 2009. Effect of TiO$_2$ nanoparticles on the surface morphology and performance of microporous PES membrane. *Appl. Surf. Sci.* 255: 4725–4732.
40. Oh, S.J., Kim, N., Lee, Y.T. 2009. Preparation and characterization of PVDF/TiO$_2$ organic-inorganic composite membranes for fouling resistance improvement. *J. Membr. Sci.* 345: 13–20.
41. Razmjou, A., Mansouri, J., Chen, V. 2011. The effects of mechanical and chemical modification of TiO$_2$ nanoparticles on the surface chemistry, structure and fouling performance of PES ultrafiltration membranes. *J. Membr. Sci.* 378: 73–84.
42. Madaeni, S.S., Zinadini, S., Vatanpour, V. 2011. A new approach to improve antifouling property of PVDF membrane using in situ polymerization of PAA functionalized TiO$_2$ nanoparticles. *J. Membr. Sci.* 380: 155–162.
43. Sotto, A., Boromand, A., S. Balta, Kim, J., Van Der Bruggen, B. 2011. Doping of polyethersulfone nanofiltration membranes: Antifouling effect observed at ultralow concentrations of TiO$_2$ nanoparticles, *J. Mater. Chem.* 21: 10311–10320.
44. Hamid, N.A.A., Ismail, A.F., Matsuura, T., Zularisam, A.W., Lau, W.J., Yuliwati, E., Abdullah, M.S. 2011. Morphological and separation performance study of polysulfone/titanium dioxide (PSF/TiO$_2$) ultrafiltration membranes for humic acid removal. *Desalination* 273: 85–92.
45. Abedini, R., Mousavi, S.M., Aminzadeh, R. 2011. A novel cellulose acetate (CA) membrane using TiO$_2$ nanoparticles: Preparation, characterization and permeation study. *Desalination* 277: 40–45.
46. Yuliwati, E., Ismail, A.F. 2011. Effect of additives concentration on the surface properties and performance of PVDF ultrafiltration membranes for refinery produced wastewater treatment. *Desalination* 273: 226–234.

47. Rahimpour, A., Jahanshahi, M., Rajaeian, B., Rahimnejad, M. 2011. TiO_2 entrapped nano-composite PVDF/SPES membranes: Preparation, characterization, antifouling and antibacterial properties. *Desalination*, 278: 343–353.

48. Sotto, A., Boromand, A., R. Zhang, Luis, P., Arsuaga, J.M., Kim, J., Van der Bruggen, B. 2011. Effect of nanoparticle aggregation at low concentrations of TiO_2 on the hydrophilicity, morphology, and fouling resistance of PES-TiO_2 membranes. *J. Colloid Interface Sci.* 363: 540–550.

49. Vatanpour, V., Madaeni, S.S., Khataee, A.R., Salehi, E., Zinadini, S., Monfared, H.A. 2012. TiO_2 embedded mixed matrix PES nanocomposite membranes: Influence of different sizes and types of nanoparticles on antifouling and performance. *Desalination*, 292: 19–29.

50. Razmjou, A., Resosudarmo, A., Holmes, R.L., Li, H., Mansouri, J., Chen, V. 2012. The effect of modified TiO_2 nanoparticles on the polyethersulfone ultrafiltration hollow fiber membranes. *Desalination*, 287: 271–280.

51. Homaeigohar, S.S., Mahdavi, H., Elbahri, M. 2012. Extraordinarily water permeable sol-gel formed nanocomposite nanofibrous membranes. *J. Colloid Interface Sci.*, 366: 51–56.

52. Zhao, S., Wang, P., Wang, C., Sun, X., Zhang, L. 2012. Thermostable PPESK/TiO_2 nanocomposite ultrafiltration membrane for high temperature condensed water treatment. *Desalination* 299: 35–43.

53. Abedini, R., Mousavi, S.M., Aminzadeh, R. 2012. Effect of sonochemical synthesized TiO_2 nanoparticles and coagulation bath temperature on morphology, thermal stability and pure water flux of asymmetric cellulose acetate nanocomposite membranes prepared via phase inversion method. *Chem. Ind. Chem. Eng. Q.* 18: 385–398.

54. Vatanpour, V., Madaeni, S.S., Moradian, R., Zinadini, S., Astinchap, B. 2012. Novel antibiofouling nanofiltration polyethersulfone membrane fabricated from embedding TiO_2 coated multiwalled carbon nanotubes. *Sep. Purif. Technol.* 90: 69–82.

55. Ngang, H.P., Ahmad, A.L., Low, S.C., Ooi, B.S. 2012. Preparation of mixed-matrix membranes for micellar enhanced ultrafiltration based on response surface methodology. *Desalination* 293: 7–20.

56. Zhang, G., Lu, S., Zhang, L., Meng, Q., Shen, C., Zhang, J. 2013. Novel polysulfone hybrid ultrafiltration membrane prepared with TiO_2-g-HEMA and its antifouling characteristics. *J. Membr. Sci.* 436: 163–173.

57. Zhang, F., Zhang, W., Yu, Y. Deng, B., Li, J., Jin, J. 2013. Sol-gel preparation of PAA-g-PVDF/TiO_2 nanocomposite hollow fiber membranes with extremely high water flux and improved antifouling property. *J. Membr. Sci.* 432: 25–32.

58. Teli, S.B., Molina, S., Sotto, A., Calvo, E.G., Abajo, J.D. 2013. Fouling resistant polysulfone-PANI/TiO_2 ultrafiltration nanocomposite membranes. *Ind. Eng. Chem. Res.* 52: 9470–9479.

59. Teow, Y.H., Ahmad, A.L., Lim, J.K., Ooi, B.S. 2013. Studies on the surface properties of mixed-matrix membrane and its antifouling properties for humic acid removal. *J. Appl. Polym. Sci.* 128: 3184–3192.

60. Joo Kim, H., Pant, H.R., Hee Kim, J., Jung Choi, N., Kim, C.S. 2014. Fabrication of multifunctional TiO_2-fly ash/polyurethane nanocomposite membrane via electrospinning. *Ceram. Int.* 40: 3023–3029.

61. Rahimpour, A., Madaeni, S.S., Taheri, A.H., Mansourpanah, Y. 2008. Coupling TiO_2 nanoparticles with UV irradiation for modification of polyethersulfone ultrafiltration membranes. *J. Membr. Sci.* 313: 158–169.

62. Damodar, R.A., You, S.J., Chou, H.H. 2009. Study the self cleaning, antibacterial and photocatalytic properties of TiO_2 entrapped PVDF membranes. *J. Hazard. Mater.* 172: 1321–1328.

63. Ngang, H.P., Ooi, B.S., Ahmad, A.L., Lai, S.O. 2012. Preparation of PVDF-TiO$_2$ mixed-matrix membrane and its evaluation on dye adsorption and UV-cleaning properties. *Chem. Eng. J.* 197: 359–367.

64. Sun, M., Su, Y., Mu, C., Jiang, Z. 2010. Improved antifouling property of PES ultrafiltration membranes using additive of silica-PVP nanocomposite. *Indust. Eng. Chem. Res.* 49: 790–796.

65. Shen, J.N., Ruan, H.M., Wu, L.G., Gao, C.J. 2011. Preparation and characterization of PES-SiO$_2$ organic inorganic composite ultrafiltration membrane for raw water pretreatment. *Chem. Eng. J.* 168: 1272–1278.

66. Muriithi, B., Loy, D.A. 2012. Processing, morphology, and water uptake of nafion/Ex situ stöber silica nanocomposite membranes as a function of particle size. *ACS Appl. Mater. Interf.* 4: 6766–6773.

67. Huang, J., Zhang, K., Wang, K., Xie, Z., Ladewig, B., Wang, H. 2012. Fabrication of polyethersulfone mesoporous silica nanocomposite ultrafiltration membranes with antifouling properties. *J. Membr. Sci.* 423–424: 362–370.

68. Pakizeh, M., Moghadam, A.N., Omidkhah, M.R., Namvar-Mahboub, M. 2013. Preparation and characterization of dimethyldichlorosilane modified SiO2/PSf nanocomposite membrane. *Korean J. Chem. Eng.* 30: 751–760.

69. Wu, H., Tang, B., Wu, P. 2014. Development of novel SiO$_2$-GO nanohybrid/polysulfone membrane with enhanced performance. *J. Membr. Sci.* 451: 94–102.

70. Yan, L., Hong, S., Li, M.L., Li, Y.S. 2009. Application of the Al$_2$O$_3$-PVDF nanocomposite tubular ultrafiltration (UF) membrane for oily wastewater treatment and its antifouling research. *Sep. Purif. Technol.* 66: 347–352.

71. Maximous, N., Nakhla, G.W., Wong, K. 2009. Preparation, characterization and performance of Al$_2$O$_3$/PES membrane for wastewater filtration. *J. Membr. Sci.* 341: 67–75.

72. Maximous, N., Nakhla, G., Wong, K., W. Wan 2010. Optimization of Al$_2$O$_3$/PES membranes for wastewater filtration. *Sep. Purif. Technol.* 73: 294–301.

73. Daraei, P., Madaeni, S.S., Ghaemi, N., Salehi, E., Khadivi, M.A., Moradian, R., Astinchap, B.. 2012. Novel polyethersulfone nanocomposite membrane prepared by PANI/Fe$_3$O$_4$ nanoparticles with enhanced performance for Cu(II) removal from water. *J. Membr. Sci.* 415–416: 250–259.

74. Gholami, A., Moghadassi, A.R., Hosseini, S.M., Shabani, S., Gholami, F. 2013. Preparation and characterization of polyvinyl chloride based nanocomposite nanofiltration-membrane modified by iron oxide nanoparticles for lead removal from water. *J. Indust. Eng. Chem,* 41:2545–2552.

75. Alam, J., Dass, L.A., Ghasemi, M., Alhoshan, M. 2013. Synthesis and optimization of PES-Fe$_3$O$_4$ mixed matrix nanocomposite membrane: Application studies in water purification. *Polymer Comp.* 34: 1870–1877.

76. Daraei, P., Madaeni, S.S., N. Ghaemi, Khadivi, M.A., Astinchap, B., Moradian, R. 2013. Fouling resistant mixed matrix polyethersulfone membranes blended with magnetic nanoparticles: Study of magnetic field induced casting. *Sep. Purif. Technol.* 109: 111–121.

77. Gohari, R.J., Lau, W.J., Matsuura, T., Halakoo, E., Ismail, A.F. 2013. Adsorptive removal of Pb(II) from aqueous solution by novel PES/HMO ultrafiltration mixed matrix membrane. *Sep. Purif. Technol.* 120: 59–68.

78. Jamshidi Gohari, R., Halakoo, E., Nazri, N.A.M., Lau, W.J., Matsuura, T., Ismail, A.F. 2014. Improving performance and antifouling capability of PES UF membranes via blending with highly hydrophilic hydrous manganese dioxide nanoparticles. *Desalination,* 335: 87–95.

79. Alhoshan, M., Alam, J., Dass, L.A., Al-Homaidi, N. 2013. Fabrication of polysulfone/ ZnO membrane: Influence of ZnO nanoparticles on membrane characteristics. *Adv. Polym. Technol.* 32: Article ID21369

80. Huang, J., Zhang, K., Wang, K., Xie, Z., Ladewig, B., Wang. H. 2012. Fabrication of polyethersulfone-mesoporous silica nanocomposite ultrafiltration membranes with anti fouling properties. *J. Membr. Sci.* 423–424: 362–370

81. Sen, D., Ghosh, A.K., Mazumder Bindal, R.C., Tewari, P.K. 2014. Novel polysulfone–spray-dried silica composite membrane for water purification: Preparation, characterization and performance evaluation. *Sep. Purif. Technol.* 123: 79–86

82. Razmjou, A., Resosudarmo, A., Holmes, R.L., Li, H., Mansouri, J., Chen, V. 2012. The effect of modified TiO_2 nanoparticles on the polyethersulfone ultrafiltration hollow fiber membranes. *Desalination* 287: 271–280

83. Rahimpour, A., Jahanshahi, M., Rajaeian, B., Rahimnejad, M. 2011.TiO_2 entrapped nano-composite PVDF/SPES membranes: Preparation, characterization, antifouling and antibacterial properties. *Desalination* 278: 343–353

84. Madaeni, S.S., Khataee, A.R., Salehi, E., Zinadini, S., Monfared, H.A., Vatanpour, V. 2012. TiO_2 embedded mixed matrix PES nanocomposite membranes: Influence of different sizes and types of nanoparticles on antifouling and performance. *Desalination* 292: 19–29

85. Zhao, S., Wang, P., Wang, C., Sun, X., Zhang, L. 2012. Thermostable PPESK/TiO_2 nanocomposite ultrafiltration membrane for high temperature condensed water treatment *Desalination* 299: 35–43

86. Hwang, H.-Y., Kim, D.-J., Kim, H.-J, Hong, Y.T., Nam, S.-Y. 2011. Effect of nanoclay on properties of porous PVDF membranes, *Trans. Nonferrous Met. Soc.* 21: 141–147.

87. Lai, C.Y., Groth, A., Gray, S., Duke, M. 2011. Investigation of the dispersion of nanoclays into PVDF for enhancement of physical membrane properties, *Desalin. Water Treat.* 34: 251–256.

88. Peng, Q.-Y, Cong, P.-H., Liu, X.J., Liu, T.X., Huang, S., Li, T.S. 2009. The preparation of PVDF/clay nanocomposites and the investigation of their tribological properties, *Wear* 266: 713–720.

89. Dayma, N., Satapathy, B.K., Patnaik, A. 2011. Structural correlations to sliding wear performance ofPA-6/PP-g-MA/nanoclay ternary nanocomposites, *Wear* 271: 827–836.

90. Pan, B., Xing, Y., Zhang, C., Zhang, Y. 2010. Study on erosion wear behavior of PDCPD/MMT nanocomposite, *Adv. Mater.Res.* 123–125: 231–234.

91. Lai, C.Y., Groth, A., Gray, S., Duke, M. 2014. Enhanced abrasion resistant PVDF/nanoclay hollow fibre composite membranes for water treatment. *J. Membr. Sci.* 449: 146–157

92. Monticelli, O., Bottino, A., Scandale, I., Capannelli, G., Russo, S. 2007. Preparation and properties of polysulfone-clay composite membranes. *J. Appl. Polym. Sci.* 103: 3637–3644.

93. Leite, A.M.D., Maia, A.M.D., Araujo, E.M., Lira, E.M. 2009. Nylon 6/Brazilian clay membranes prepared by phase inversion. *J. Appl. Polym. Sci.* 113: 1488–1493.

94. Anadão, P., Sato, L.F., Wiebeck, H., Valenzuela-Díaz, F.R. 2010. Montmorillonite as a component of polysulfone nanocomposite membranes. *Appl. Clay Sci.* 48: 127–132.

95. Lai, C.Y., Groth, A., Gray, S., Duke, M. 2011. Investigation of the dispersion of nanoclays into PVDF for enhancement of physical membrane properties. *Desalin. Water Treat.* 34: 251–256.

96. Ghaemi, N., Madaeni, S.S., Alizadeh, A., Rajabi, H., Daraei, P. 2011. Preparation, characterization and performance of polyethersulfone/organically modified montmorillonite nanocomposite membranes in removal of pesticides. *J. Membr. Sci.* 382: 135–147.

97. Ma, Y., Shi, F., Zhao, W., Wu, M., Zhang, J., Ma, J., Gao, C. 2012. Preparation and characterization of PSf/clay nanocomposite membranes with LiCl as a pore forming additive. *Desalination* 303: 39–47.

98. Wang, P., Ma, J., Wang, Z., Shi, F., Liu, Q. 2012. Enhanced separation performance of PVDF/PVP-g-MMT nanocomposite ultrafiltration membrane based on the NVP-grafted polymerization modification of montmorillonite (MMT). *Langmuir* 28: 4776–4786.

99. Anadão, P., Montes, R.R., Larocca, N.M., Pessan, L.A. 2013 Influence of the clay content and the polysulfone.molar mass on nanocomposite membrane properties. *Appl. Surf. Sci.* 275:110–120

100. Adams, F.V., Nxumalo, E.N., Krause, R.W.M., Hoek, E.M.V., Mamba, B.B. 2012. Preparation and characterization of polysulfone/β-cyclodextrin polyurethane composite nanofiltration membranes. *J. Membr. Sci.* 405–406: 291–299.

101. Adams, F.V., Dlamini, D.S., Nxumalo, E.N., R.W.M. Krause, Hoek, E.M.V., Mamba, B.B. 2013. Solute transport and structural properties of polysulfone/β-cyclodextrin polyurethane mixed-matrix membranes. *J. Membr. Sci.* 429: 58–65.

102. Adams, F.V., Nxumalo, E.N., Krause, R.W.M., Hoek, E.M.V., Mamba, B.B. 2013. The influence of solvent properties on the performance of polysulfone/β-cyclodextrin polyurethane mixed-matrix membranes. *J. Appl. Polym. Sci.* 130: 2005–2014.

103. Adams, F.V., Nxumalo, E.N., Krause, R.W.M., Hoek, E.M.V., Mamba, B.B. 2013. Application of polysulfone/cyclodextrin mixed-matrix membranes in the removal of natural organic matter from water. *Phys. Chem. Earth*, Parts A/B/C 67–69.

104. Fan, Z., Wang, Z., Sun, N., Wang, J., Wang, S. 2008. Performance improvement of polysulfone ultrafiltration membrane by blending with polyaniline nanofibers. *J. Membr. Sci.* 320: 363–371.

105. Sankir, M., Bozkir, S., Aran, B. 2010. Preparation and performance analysis of novel nanocomposite copolymer membranes for Cr(VI) removal from aqueous solutions. *Desalination* 251: 176–180.

106. Zhao, S., Wang, Z., Wei, X., Tian, X., Wang, J., Yang, S., Wang, S. 2011. Comparison study of the effect of PVP and PANI nanofibers additives on membrane formation mechanism, structure and performance. *J. Membr. Sci.* 385–386: 110–122.

107. Zhao, S., Wang, Z., Wang, J., Yang, S., Wang, S. 2011. PSf/PANI nanocomposite membrane prepared by in situ blending of PSf and PANI/NMP. *J. Membr. Sci.* 376: 83–95.

108. Teli, S.B., Molina, S., Calvo, E.G., Lozano, A.E., de Abajo, J. 2012. Preparation, characterization and antifouling property of polyethersulfone-PANI/PMA ultrafiltration membranes. *Desalination* 299: 113–122.

109. Zhao, S., Wang, Z., Wei, X., Zhao, B., Wang, J., Yang, S., Wang, S. 2012. Performance improvement of polysulfone ultrafiltration membrane using well-dispersed polyaniline-poly(vinylpyrrolidone) nanocomposite as the additive. *Indust. Eng. Chem. Res.* 51: 4661–4672.

110. Liao, Y., Wang, X., Qian, W., Li, Y., Li, X., Yu, D.G. 2012. Bulk synthesis, optimization, and characterization of highly dispersible polypyrrole nanoparticles toward protein separation using nanocomposite membranes. *J. Colloid Interface Sci.* 386: 148–157.

111. Tetala, K.K.R., Stamatialis, D.F. 2013. Mixed matrix membranes for efficient adsorption of copper ions from aqueous solutions. *Sep. Purif. Technol.* 104: 214–220.

112. Worthley, C.H., Constantopoulos, K.T., Ginic-Markovic, M., Markovic, E., Clarke, S. 2013. A study into the effect of POSS nanoparticles on cellulose acetate membranes. *J.f Membr. Sci.* 431: 62–71.

113. Zhao, W., Huang, J., Fang, B., Nie, S., Yi, N., Su, B., Li, H., Zhao, C. 2011. Modification of polyethersulfone membrane by blending semi-interpenetrating network polymeric nanoparticles. *J. Membr. Sci.* 369: 258–266.

114. Tomalia, D.A., Naylor, A.N., Goddard, W.A. 1990. Starburst dendrimers: Molecular level control of size, shape, surface, chemistry and topology. *Angew. Chem. Int. Ed. Engl.* 29: 138–175.

115. Fréchet, J.M.J. 1994. Functional polymers and dendrimers: Reactivity, molecular architecture, and interfacial energy. *Science* 263: 1710–1715.

116. Fréchet, J.M.J., Tomalia, D.A. 2001. *Dendrimers and Other Dendritic Polymers*, Wiley-VCH, Weinheim.

117. Cheng, Y. 2012. *Dendrimer Based in Drug Delivery Systems*, John Wiley &Sons, Hoboken, NJ.
118. Moorefield, C.N, Perera, S., Newkome, G.R. 2012. Chapter 1: Dendrimer chemistry: Supramolecular perspectives and applications, in: Y. Cheng (Ed.), *Dendrimer Based Drug Delivery Systems: From Theory to Practice*, John Wiley &Sons, Hoboken, NJ.
119. Boas, U., Heegaard, P.M. 2004. Dendrimers in drug research, *Chem. Soc. Rev.* 33: 43–63.
120. Algarra, M., Campos, B.B., Miranda, M.S., Esteves da Silva, J.C.G. 2011. CdSequantumdots capped PAMAM dendrimer nanocomposites for sensing nitroaromatic compounds, *Talanta* 83: 1335–1340.
121. Esteves da Silva, J.C.G., Algarra, M., Campos, B.B. 2011. Synthesis and analytical applications of Quantum Dots coated with different generations of DABdendrimers, in: B. Reddy (Ed.), *Advances in Nanocomposites-Synthesis, Characterization and Industrial Applications*, IN-TECH, London, 23–38.
122. Algarra, M., Campos, B.B., Alonso, B., Miranda, M.S., Martínez, A.M., Casado, C.M., Esteves da Silva, J.C.G. 2012. Thiolated DAB dendrimers and CdSe quantum dots nanocomposites for Cd(II) or Pb(II) sensing. *Talanta* 88: 403–407.
123. Jie, G., Yuan, J. 2012. Quantum dots-based multifunctional dendritic superstructurefor amplified electrochemiluminescence detection of ATP. *Biosens. Bioelectron.* 31: 69–76.
124. Oliveira, J.M., Salgado, A.J., Sousa, N., Mano, J.F., Reisa, R.L. 2010. Dendrimers and derivatives as a potential therapeutic tool in regenerative medicine strategies–a review, *Prog. Polym. Sci.* 35: 1163–1194.
125. Vogtle, F., Richardt, G., Werner, N. 2009. *Dendrimer Chemistry: Concepts, Syntheses, Properties, Applications*, Wiley-VCH, Weinheim.
126. Rether, A., Schuster, M. 2003. Selective separation and recovery of heavy metal ions using water-soluble N-benzoylthiourea modified PAMAM polymers. *React. Funct. Polym.* 57: 13–21.
127. Diallo, M.S., Christie, S., Swaminathan, P., Balogh, L., Shi, X., Um, W., Papelis, C., Goddard, W.A., Johnson, J.H. 2004. Dendritic chelating agents. 1. Cu(II) binding to ethylenediamine core poly(amidoamine) dendrimers in aqueous solutions. *Langmuir* 20: 2640–2651.
128. Diallo, M.S., Christie, S., Swaminathan, P., Johnson, J.H., Goddard, W. 2005. Dendrimer enhanced ultrafiltration. 1. Recovery of Cu(II) from aqueous solutions using PAMAM dendrimers with ethylenediamine core and terminal NH_2 groups, *Environ. Sci. Technol.* 39: 1366–1377.
129. Diallo, M.S., Arasho, W., Johnson, J.H., Goddard, W.A. 2008. Dendritic chelating agents U(VI) binding to poly(amidoamine) and poly(propyleneimine)dendrimers in aqueous solutions. *Environ. Sci. Technol.* 42: 1572–1579.
130. Xu, Y., Zhao, D. 2005. Removal of Cu (II) from contaminated soil using PAMAM dendrimers. *Environ. Sci. Technol.* 39: 2369–2377.
131. Barakat, M.A., Ramadan, M.H., Alghamadi, M.A., Algarny, S.S., Woodcock, H.L., Kuhn, J. 2013. Remediation of Cu (II), Ni(II), and Cr(III) ions from simulated wastewater by dendrimer/titania composites. *J. Environ. Manage.* 117: 50–57.
132. Algarra, M., Vázquez, A.M.I., Alonso, B., Casado, C.M., Casado, J., Benavente, J. 2014. Characterization of an engineered cellulose based membrane by thiol dendrimer for heavy metals removal. *Chem. Eng. J.* 253: 472–477.
133. Husain, S., Koros, W.J. 2009. Macro Voids in hybrid organic/inorganic hollow fiber membranes. *Indust. Eng. Chem. Res.* 48: 2372–2379.
134. Junaidi, M.U.M., Leo, C.P., Kamal, S.N.M., Ahmad, A.L. 2013. Fouling mitigation in humic acid ultrafiltration using polysulfone/SAPO-34 mixed matrix membrane. *Water Sci. Technol.* 67: 2102-2109.
135. Leo, C.P., Ahmad Kamil, N.H., Junaidi, M.U.M., Kamal, S.N.M., Ahmad, A.L. 2013. The potential of SAPO-44 zeolite filler in fouling mitigation of polysulfone ultrafiltration membrane. *Sep. Purif. Technol.* 103: 84–91.

136. Liu, F., Ma, B.R., Zhou, D., Xiang, Y.H., Xue, L.X. 2014. Breaking through tradeoff of Polysulfone ultrafiltration membranes by zeolite 4A. *Microporous Mesoporous Mater.* 186: 113–120.

137. Ciobanu, G., Carja, G., Ciobanu, O. 2007. Preparation and characterization of polymer-zeolite nanocomposite membranes. *Mater. Sci. Eng. C* 27: 1138–1140.

138. Liao, C., Yu, P., Zhao, J., Wang, L., Luo, Y. 2011. Preparation and characterization of NaY/PVDF hybrid ultrafiltration membranes containing silver ions as antibacterial materials. *Desalination* 272: 59–65.

139. Sotiriou, G.A., Pratsinis, S.E. 2010. Antibacterial activity of nanosil versions and particles. *Environ. Sci. Technol.* 44: 5649–5654.

140. Levard, C., Hotze, E.M., Lowry, G.V., Brown, G.E. 2012. Environmental transformations of silver nanoparticles: Impact on stability and toxicity. *Environ. Sci. Technol.* 466: 900–6914.

141. Koseoglu-Imer, D.Y., Kose, B., Altinbas, M., Koyuncu, I. 2013. The production of polysulfone(PS) membrane with silver nanoparticles (AgNP): Physical properties, filtration performances, and biofouling resistance of membranes. *J. Membr. Sci.* 428: 620–628.

142. Zodrow, K., Brunet, L.A., Mahendra, S.D., Zhang, A., Li, Q., Alvarez, P.J.J. 2009. Polysulfone ultrafiltration membranes impregnated with silver nanoparticles show improved biofouling resistance and virus removal. *Water Res.* 43: 715–723.

143. Sawada, I., Fachrul, R., Ito, T., Ohmukai, Y., Maruyama, T., Matsuyama, H. 2012. Development of a hydrophilic polymer membrane containing silver nanoparticles with both organic antifouling and antibacterial properties. *J. Membr. Sci.* 387–388: 1–6.

144. Basri, H., Ismail, A.F., Aziz, M. 2011. Polyethersulfone (PES)-silvercomposite UF membrane: Effect of silver loading and PVP molecular weight on membrane morphology and antibacterial activity, *Desalination* 273: 72–80.

145. Basri, H., Ismail, A.F., Aziz, M., Nagai, K., Matsuura, T., Abdullah, M.S., Ng, B.C. 2010 Silver-filled polyethersulfone membranes for antibacterial applications — Effect of PVP and TAP addition on silver dispersion. *Desalination* 261: 264–271.

146. Huang, J., Arthanareeswaran, G., Zhang, K. 2012. Effect of silver loaded sodium zirconium phosphate (nanoAgZ) nanoparticles incorporation on PES membrane performance. *Desalination* 285: 100–107.

147. Li, J.H., Shao, X.S., Zhou, Q., Li, M.Z., Zhang, Q.Q. 2013. The double effects of silver nanoparticles on the PVDF membrane: Surface hydrophilicity and antifouling performance, *Appl. Surf. Sci.* 265: 663–670.

148. De Gusseme, B., Hennebel, T., Christiaens, E., Saveyn, H., Verbeken, K., Fitts, J.P., Boon, N., Verstraete, W. 2011. Virus disinfection in water by biogenic silver immobilized in polyvinylidene fluoride membranes. *Water Res.* 45: 1856–1864.

149. Yin, J., Yang, Y., Hu, Z., Deng, B. 2013. Attachment of silver nanoparticles (AgNPs) onto thin-film composite (TFC) membranes through covalent bonding to reduce membrane biofouling. *J. Membr. Sci.* 441: 73–82.

150. Lee, S.Y., Kim, H.J., Patel, R., Im, S.J., Kim, J.H., Min, B.R. 2007.Silver nanoparticles immobilized on thin film composite polyamide membrane: Characterization, nanofiltration, antifouling properties. *Polym. Adv. Technol.* 18: 562–568.

151. Regiel, A., Irusta, S., Kyziol, A., Arruebo, M., Santamaria, J. 2013. Preparation and characterization of chitosan-silver nanocomposite films and the antibacterial activity against Staphylococcusaureus. *Nanotechnology* 24.

152. Mollahosseini, A., Rahimpour A., Jahamshahi, M., Peyravi, M., Khavarpour, M. 2012. The effect of silver nanoparticle size on performance and antibacterial of polysulfone ultrafiltration membrane. *Desalination*, 306: 41–50.

153. Zhang, M., Field, R.W., Zhang, K. 2014. Biogenic silver nanocomposite polyether sulfone UF membranes with antifouling properties. *J. Membr. Sci.* 471: 274–284.

154. Sawada, I., Fachrul, R., Ito, T., Yoshikage, O., Tatsuo, M., Matsuyama, H. 2012. Development of a hydrophilic polymer membrane containing silver nanoparticles with both organic antifouling and antibacterial properties. *J. Memb. Sci.* 387– 388: 1–6.
155. Alpatova, A., Sikkim, E., Sun, X., Hwang, G., Yang, L., Gamal El-Din, M. 2013. Fabrication of porous polymeric nanocomposite membranes with enhanced anti-fouling properties: Effect of casting composition. *J. Membr. Sci.* 444: 449–460.
156. Kar, S., Subramanian, M., Ghosh, A.K., Bindal, R.C., Prabhakar, S., Nuwad, J., Pillai, S. Chattopadhyay, C.G.S., Tewari, P.K. 2011. Potential of nanoparticles for water purification: A case-study on anti-biofouling behaviour of metal based polymeric nanocomposite membrane. *Desalinat. Water Treat.*, 27: 224–230.
157. Dasari, A., Quirós, J., Herrero, B., Boltes, K., García-Calvo, E., Rosal, R. 2012. Antifouling membranes prepared by electrospinning polylactic acid containing biocidal nanoparticles. *J. Membr. Sci.* 405–406: 134–140.
158. Akar, N., Asar, B., Dizge, N., Koyuncu, I. 2013. Investigation of characterization and biofouling properties of PES membrane containing selenium and copper nanoparticles. *J. Membr. Sci.* 437: 216–226.
159. Szabó, T., Tombácz, E., Illés, E., Dékány, I. 2006. Enhanced acidity and pH-dependent surface charge characterization of successively oxidized graphite oxides. *Carbon* 44: 537–545.
160. Ganesh , B.M., Isloor A.M., Ismail, A.F. 2013. Enhanced hydrophilicity and salt rejection study of graphene oxide-polysulfone mixed matrix membrane. *Desalination* 313: 199–207.
161. Zhao, C., Xu, X., Chen, J., Yang, F. 2013. Effect of graphene oxide concentration on the morphologies and antifouling properties of PVDF ultrafiltration membranes. *J. Environ. Chem. Eng.* 1: 349–354.
162. Jin, F., Wei, L., Chen, Z, Li, Z., Rongxin, S., Wei, Q., Yang, Q.H., and He, Z.. 2013. High-performance ultrafiltration membranes based onpolyethersulfone–graphene oxide composites. *RSC Adv.* 3: 21394–21397.
163. Wang, Z., Yu, H., Xia, J., Zhang, F., Li, F., Xia, Y., Li, Y. 2012. Novel GO-blended PVDF ultrafiltration membranes. *Desalination* 299: 50–54.
164. Le-Clech, P., Chen, V., Fane, T.A.G. 2006. Fouling in membrane bioreactors used in wastewater treatment. *J. Membr. Sci.* 284: 17–53.
165. Meng, F., Chae, S.-R., Drews, A., Kraume, M., Shin, H.S., Yang, F. 2009. Recent advances in membrane bioreactors (MBRs): Membrane fouling and membrane material. *Water Res.* 43: 1489–1512.
166. Zhang, H., Gao, J., Jiang, T., Gao, D., Zhang, S., Li, H., Yang, F. 2011. A novel approach to evaluate the permeability of cake layer during cross-flow filtration in the flocculants added membrane bioreactors. *Bioresour. Technol.* 102: 11121–11131.
167. Wu, B., Kitade, T., Chong, T.H., Uemura, T., Fane, A.G. 2012. Role of initially formed cakelayers on limiting membrane fouling in membrane bioreactors. *Bioresour. Technol.* 118: 589–593.
168. Meng, F., Zhang, H., Yang, F., Liu, L. 2007. Characterization of cake layer in submerged membrane bioreactor. *Environ. Sci. Technol.* 41: 4065–4070.
169. Zhao, C., Xu, X., Chena, J., Wang, G., Yang, F. 2014. Highly effective antifouling performance of PVDF/graphene oxide composite membrane in membrane bioreactor (MBR) system. *Desalination* 340: 59–66.
170. Vatanpour, V., Madaeni, S.S., Moradian, R., Zinadini, S., Astinchap, B. 2012. Novel anti biofouling nanofiltration polyethersulfone membrane fabricated from embedding TiO$_2$ coated multiwalled carbon nanotubes. *Sep. Purif. Technol.*, 90: 69–82.
171. Crock, C.A., Rogensues, A.R., Shan, W., Tarabara, V.V. 2103. Polymer nanocomposites with graphene based hierarchical fillers as materials for multifunctional water treatment membranes. *Water Res.* 47: 3984–3996.

172. Wu, H., Tang, B., Wu, P. 2014. Development of novel SiO2-GO nanohybrid/polysulfone membrane with enhanced performance. *J. Membr. Sci.* 451: 94–102.

173. Filice, S., D'Angelo, D., Libertino, S., Nicotera, I., Kosma, V., Privitera, V., Scalese, S. 2015. Graphene oxide and titania hybrid Nafion membranes for efficient removal of methyl orange dye from water. *Carbon* 8: 489–499.

174. Safarpoura, M., Khataee, A., Vatanpour, V. 2015. Effect of reduced graphene oxide/TiO$_2$ nanocomposite with different molar ratios on the performance of PVDF ultrafiltration membranes. *Sep. Purif. Technol.* 140: 32–42.

175. Li, G., Shen, L., Luo, Y., Zhang, S. 2014. The effect of silver-PAMAM dendrimer nanocomposites on the performance of PVDF membranes. *Desalination* 338: 115–120.

176. Cadotte, J.E. 1981. *Interfacially synthesized reverse osmosis membrane.* US Patent No. US4277344.

177. Jeong, B.-H., Hoek, E.M.V., Yan, Y., Subramani, A., Huang, X., Hurwitz, G., Ghosh, A.K., Jawor, A. 2007. Interfacial polymerization of thin film nanocomposites: A new concept for reverse osmosis membranes. *J. Membr. Sci.* 294: 1–7.

178. Glater, J., Hong, S.k., Elimelech, M. 1994.The search for a chlorine-resistant reverse osmosis membrane. *Desalination* 95: 325–345.

179. Kwon, Y.N., Hong, S., Choi, H., Tak, T. 2012. Surface modification of a polyamide reverse osmosis membrane for chlorine resistance improvement. *J. Membr. Sci.* 415–416: 192–198.

180. Shintani, T., Matsuyama, H., Kurata, N. 2007. Development of a chlorine-resistant polyamide reverse osmosis membrane. *Desalination* 207: 340–348.

181. Park, J., Choi, W., Kim, S.H., Chun, B.H., Bang, J., Lee, K.B. 2010. Enhancement of chlorine resistance in carbon nanotube-based nanocomposite reverse osmosis membranes. *Desalin. Water Treat.* 15: 198–204.

182. Zhao, H., Qiu, S., Wu, L., Zhang, L., Chen, H., Gao, C. 2014. Improving the performance of polyamide reverse osmosis membrane by incorporation of modified multi-walled carbon nanotubes. *J. Membr. Sci.* 450: 249–256.

183. Kim, S.G., Hyeon, D.H., Chun, J.H., Chun, B.H., Kim, S.H. 2013. Nanocomposite poly(arylene ether sulfone) reverse osmosis membrane containing functional zeolite nanoparticles for seawater desalination. *J. Membr. Sci.* 443: 10–18.

184. Kim, S.G., Chun, J.H., Chun, B.H., Kim, S.H. 2013. Preparation, characterization and performance of poly(aylene ether sulfone)/modified silica nanocomposite reverse osmosis membrane for seawater desalination. *Desalination* 325: 76–83.

185. Jadav, G.L., Singh, P.S. 2009. Synthesis of novel silica-polyamide nanocomposite membrane with enhanced properties. *J. Membr. Sci.* 328: 257–267.

186. Jadav, G.L., Aswal, V.K., Singh, P.S. 2010. SANS study to probe nanoparticle dispersion in nanocomposite membranes of aromatic polyamide and functionalized silica nanoparticles. *J. Colloid Interf. Sci.* 351: 304–314.

187. Fathizadeh, M., Aroujalian, A., Raisi, A. 2011. Effect of added NaX nano-zeolite into polyamide as a top thin layer of membrane on water flux and salt rejection in a reverse osmosis process. *J. Membr. Sci.e* 375: 88–95.

188. Rajaeian, B., Rahimpour, A., Tade, M.O., Liu, S. 2013. Fabrication and characterization of polyamide thin film nanocomposite (TFN) nanofiltration membrane impregnated with TiO$_2$ nanoparticles. *Desalination* 313: 176–188.

189. Kim, E.-S., Deng, B. 2011. Fabrication of polyamide thin-film nano-composite (PA-TFN) membrane with hydrophilized ordered mesoporous carbon (H-OMC) for water purifications. *J. Membr. Sci.* 375: 46–54.

190. Rana, D., Kim, Y., Matsuura, T., Arafat, H.A. 2011. Development of antifouling thin-film-composite membranes for seawater desalination. *J. Membr. Sci.* 367: 110–118.

191. Lee, S.Y., Kim, H.J., Patel, R., Im, S.J., Kim, J.H., Min, B.R. 2007. Silver nanoparticles immobilized on thin film composite polyamide membrane: Characterization, nanofiltration, antifouling properties. *Polym. Adv. Technol.* 18: 562–568.

192. Kim, E.S., G. Hwang, Gamal El-Din, M., Liu, Y. 2012. Development of nanosilver and multi-walled carbon nanotubes thin-film nanocomposite membrane for enhanced water treatment. *J. Membr. Sci.* 394–395: 37–48.

193. Lind, M.L., Jeong, B.H., Subramani, A., Huang, X., Hoek, E.M.V. 2009. Effect of mobile cations on zeolite polyamide thin film nanocomposite membranes. *J. Mater. Res.* 24: 1624–1631.

194. Li, D., Wang, H. 2010. Recent developments in reverse osmosis desalination membranes. *J. Mater. Chem.* 20: 4551–4566.

195. Bellona, C., Drewes, J.E., Xu, P., Amy, G. 2004. Factors affecting the rejection of organic solutes during NF/RO treatment - A literature review. *Water Res.* 38: 2795–2809.

196. Roh, I.J., Greenberg, A.R., Khare, V.P. 2006. Synthesis and characterization of interfacially polymerized polyamide thin films. *Desalination* 191: 279–290.

197. Zhang, L., Shi, G.Z., Qiu, S., Cheng, L.H., Chen, H.L. 2011. Preparation of high-flux thin film nanocomposite reverse osmosis membranes by incorporating functionalized multi-walled carbon nanotubes. *Desalin. Water Treat.* 34: 19–24.

198. Lind, M.L., Ghosh, A.K., Jawor, A., Huang, W., Hou, Y., Hoek, E.M.V. 2009. Influence of zeolite crystal size on zeolite-polyamide thin film nanocomposite membranes. *Langmuir* 25: 10139–10145.

199. Kim, C.K., Kim, J.H., Roh, I.J., Kim, J.J. 2000. The changes of membrane performance with polyamide molecular structure in the reverse osmosis process. *J. Membr. Sci.* 165: 189–199.

200. Ghosh, A.K., Jeong, B.H., Huang, X., Hoek, E.M.V. 2008. Impacts of reaction and curing conditions on polyamide composite reverse osmosis membrane properties. *J. Membr. Sci.* 311: 34–45.

201. Lind, M.L., Suk, D.E., Nguyen, T.V., Hoek, E.M.V. 2010. Tailoring the structure of thin film nanocomposite membranes to achieve seawater RO membrane performance. *Environ. Sci. Technol.* 44: 8230–8235.

202. Roy, S., Ntim Mitra, S., Sirkar, K.K. 2011. Facile fabrication of superior nanofiltration membranes from interfacially polymerized CNT-polymer composites. *J. Membr. Sci.* 375: 81–87.

203. Pendergast, M.T.M., Nygaard, J.M., Ghosh, A.K., Hoek, E.M.V. 2010. Using nanocomposite materials technology to understand and control reverse osmosis membrane compaction, *Desalination*. 261: 255–263.

204. Kwak, S.Y., Kim, S.H., Kim, S.S. 2001. Hybrid organic/inorganic reverse osmosis (RO) membrane for bactericidal anti-fouling. 1. Preparation and characterization of TiO_2 nanoparticle self-assembled aromatic polyamide thin-film-composite (TFC) membrane. *Environ. Sci. Technol.* 35: 2388–2394.

205. Kim, S.H., Kwak, S.Y., Sohn, B.H., Park, T.H. 2003. Design of TiO_2 nanoparticle self-assembled aromatic polyamide thin-film-composite (TFC) membrane as an approach to solve biofouling problem. *J. Membr. Sci.* 211: 157–165.

206. Bae, T.H., Tak, T.M. 2005. Preparation of TiO_2 self-assembled polymeric nanocomposite membranes and examination of their fouling mitigation effects in a membrane bioreactor system. *J. Membr. Sci.* 266: 1–5.

207. Pendergast, M.T.M., Eric Hoek, M.V. 2011. A review of water treatment membrane nanotechnologies. *Energy Environ. Sci.* 4: 1946–1971.

208. Li, Y.-H., Wang, S., Wei, J., Zhang, X., Xu, C., Luan, Z., Wu, D., Wei, B. 2002. Lead adsorption on carbon nanotubes. *Chem. Phys. Lett.* 357: 263–266.

209. Li, Y.-H., Wang, S., Luan, Z., Ding, J., Xu, C., Wu, D. 2003. Adsorption of cadmium(II) from aqueous solution by surface oxidized carbon nanotubes. *Carbon* 41: 1057–1062.

210. Li, Y.-H., Wang, S., Zhang, X., Wei, J., Xu, C., Luan, Z., Wu, D. 2003. Adsorption of fluoride from water by aligned carbon nanotubes. *Mater. Res. Bull.* 38: 469–476.

211. Peng, X., Luan, Z., Ding, J., Di, Z., Li, Y., Tian, B. 2005. Ceria nanoparticles supported nanotubes for the removal of arsenate from water. *Mater. Lett.* 59: 399–403.
212. Srivastava, A., Srivastava, O.N., Talapatra, S., Vajtai, R., Ajayan, P.M. 2004. Carbon nanotube filters. *Nat. Mater.* 3: 610–614.
213. Kar, S., Bindal, R.C., Tewari, P.K., Dasgupta, K., Sathiyamoorthy, D. 2008. Potential of carbon nanotubes in water purification: An approach towards the development of an integrated membrane system. *Intl. J. Nucl. Desalin.* 3: 143–150.
214. Rosca, I.D., Watari, F., Uo, M., Akasaka, T. 2005. Oxidation of multiwalled carbon nanotubes by nitric acid. *Carbon* 15: 3124–3131.
215. Sano, M., Kamino, A., Okamura, J., Shinkai, S. 2001. Ring closure of carbon nanotubes. *Science* 293: 1299–1301.
216. Hamwi, A., Alvergnat, H., Bonnamy, S., Beguin, F. 1997. Fluorination of carbon nanotubes. *Carbon* 35: 723–728.
217. Mickelson, E.T., Huffman, C.B., Rinzler, A. G, Smalley, H., Margrave, J.L. 1998. Fluorination of single-wall carbon nanotubes. *Chem. Phys. Lett.* 296: 188–194.
218. Touhara, H., Okino, F. 2000. Property control of carbon material by fluorination. *Carbon* 38: 241–267.
219. Khare, B.N., Meyyappan, M., Cassell, A.M., Nguyen, C.V., Han, J., Nano, J. 2002. Functionalisation of carbon nanotubes using atomic hydrogen from a glow discharge. *Letters* 2: 73–77.
220. Chen, Y., Haddon, R.C., Fang, S., Rao, A.M., Eklund, P.C., Lee, W.H., Dickey, E.C., Grulke, E.A., Pendergrass, J.C., Chavan, A., Haley, B.E., Smalley, R.E. 1998.Chemical attachment of organic functional groups to single-walled carbon nanotube material, *J. Mater. Res.* 13: 2423–2431.
221. Chen, J., Hamon, M.A., Hu, H., Chen, Y., Rao, A.M., Eklund, P.C., Haddon, R.C. 1998. Solution properties of single walled carbon nanotubes, *Science* 282: 95–98.
222. Lee, W. H, Kim, S.J., Lee, W.J., Lee, J.G., Haddon, R.C., Reucroft, P.J. 2001. X-ray photoelectron spectroscopic studies of surface modified single-walled carbon nanotube material. *Appl. Surf. Sci.* 181: 121–127.
223. Kamaras, K., Itkis, M.E., Hu, H., Zhao, B., Haddon, R.C. 2003. Covalent bond formation to a carbon nanotube metal. *Science* 301: 1501.
224. Hu, H., Zhao, B., Hamon, M.A., Kamaras, K., Itkis, M.E., Haddon, R.C. 2003. Sidewall functionalization of single walled carbon nanotubes by addition of dichloro carbine. *J. Am. Chem. Soc.* 125: 14893–14900.
225. Khabashesku, V.N., Billups, W.E., Margrave, J.L. 2002. Fluorination of single-wall carbon nanotubes and subsequent derivatization reactions. *Acc. Chem. Res.* 35: 1087–1095.
226. Stevens, J.L., Huang, A.Y., Peng, H.Q., Chiang, L.W., Khabashesku, V.N. Margrave, J.L. 2003. Side wall amino-functionalization of single-walled carbon nanotubes through fluorination and subsequent reactions with terminal diamines. *Nano Lett.* 3: 331–336.
227. Dujardin, E., Ebbesen, T.W., Hiura, H., Tanigaki, K. 1994. Capillarity and wetting of carbon nanotubes. *Science* 265: 1850–1852.
228. Ajayan, P.M., Iijima, S. 1993. Capillarity induced filling of carbon nanotubes. *Nature* 361:333–334.
229. Ajayan, P.M., Ebbesen, T.W., Ichihasi, T., Iijima, S., Tanigaki, K., Hiura, H. 1993. Opening carbon nanotubes with oxygen and implications for filling. *Nature* 362: 522–525.
230. Ajayan, P.M., Stephan, O., Redlich, P., Coliex, C. 1995. Carbon nanotubes as removable templates for metal-oxide nanocomposites. *Nature* 375: 564–567.
231. Yarin, A.L., Yazicioglu, A.G, Megaridis, C.M., Rossi, M.P., Gogotsi, Y. 2005. Theoretical and experimental investigation of aqueous liquids contained in carbon nanotubes. *J. Appl. Phys.* 97: 124309/1–124309/13.

232. Sholl, D.S., Fichthorn, K.A. 1997. Concerted diffusion of molecular clusters in a molecular sieve. *Phys. Rev. Lett.*79: 3569–3572.

233. Sholl, D.S., Lee, C.K. 2000. Influences of concerted cluster diffusion on single-file diffusion of CF4 in AlPO4-5 and Xe in AlPO4-31. *J. Chem. Phys.* 112: 817–824.

234. Brovchenko, I., Geiger, A., Oleinikova, A. 2001. Phase equilibria of water in cylindrical nanopores, *Phys. Chem. Chem. Phys.* 3: 1567–1569.

235. Allen, W., Kuyucak, S., Chung, S.H. 1999. The effect of hydrophobic and hydrophilic channel walls on the structure and diffusion of water and ions, *J. Chem. Phys.* 111: 7985–7999.

236. Marti, J., Gordillo, M.C., 2002. Microscopic dynamics of confined supercritical water. *Chem. Phys. Lett.* 354: 227–232.

237. Hummer, G., Rasaiah, J.C., Noworyta, J.P. 2001. Water conduction through the hydrophobic channel of a carbon nanotube, *Nature* 414: 188–190.

238. Thomas, J.A., McGaughey, A.J.H. 2008. Reassessing fast water transport through carbon nanotubes, *Nano Lett.* 8:2788–2793.

239. Verwei, H., van den Boom, H. 2007. Fast mass transport through carbon nanotube membranes, *Small.* 3: 1996–2004.

240. Striolo, A. 2006. The mechanism of water diffusion in narrow carbon nanotubes. *Nano Lett.* 6: 633–639.

241. Kang, S., Herzberg, M., Rodrigues, D.F., Elimelech, M., 2008. Antibacterial effects of carbon nanotubes: Size does not matter! *Langmuir.* 24: 6409–6413.

242. Narayan, R.J., Berry, C.J., Brigmon, R.L. 2005. Structural and biological properties of carbon nanotube composite films. *Mater. Sci. Eng. B* 123: 123–129.

243. Wick, P., Manser, P., Limbach, L.K., Dettlaff-Weglikowska, U., Krumeich, F., Roth, S., Stark, S.W.J., Bruinink, A. 2007. The degree and kind of agglomeration affect carbon nanotube cytotoxicity. *Toxicol. Lett.* 168: 121–131.

244. Brunet, L., Lyon, D.Y., Zodrow, K., Rouch, J.C., Caussat, B., Serp P., Remigy, J.C., Wiesner, M.R., Alvarez, P.J.J. 2008. Properties of membranes containing semi dispersed carbon nanotubes, *Environ. Eng. Sci.* 25: 565–576.

245. Majumder, M., Ajayan, P.M. 2010. Carbon nanotube membranes: A new frontier in membrane science. *Comprehensive Membrane Science and Engineering*, Vol. 1, 291–310, Academic Press, Oxford.

246. Miller, S.A., Young, V.Y., Martin, C.R.J. 2001. Electroosmotic flow in template-prepared carbon nanotube membranes. *Am. Chem. Soc.*123:12335–12342.

247. Hinds, B.J., Chopra, N., Rantell, T., Andrews, R., Gavalas, V., Bachas, L.G. 2004. Aligned multiwalled carbon nanotube membranes. *Science* 303: 62–65.

248. Holt, J.K., Park, H.G., Wang, Y., Stadermann, M., Artyukhin, A.B., Grigoropoulosm, C.P., Noy, A., Bakajin, O. 2006. Fast mass transport through sub-2-nanometer carbon nanotubes. *Science* 312: 1034–1037.

249. Majumder, M., Chopra, N., Andrews, R., Hinds, B.J. 2005. Nanoscale hydrodynamics: Enhanced flow in carbon nanotubes. *Nature* 438: 44.

250. Majumder, M., Chopra, N., Hinds, B.J. 2005. Effect of tip functionalization on transport through vertically oriented carbon nanotube membranes. *J. Am. Chem. Soc.* 127: 9062–9070.

251. Kang, S., Pinault, M., Pfefferle, L.D., Elimelech, M. 2007. Single-walled carbon nanotubes exhibit strong antimicrobial activity. *Langmuir* 23: 8670–8673.

252. Brady-Este´vez, A.S., Kang, S., Elimelech, M. 2008. A single-walled-carbon-nanotube filter for removal of viral and bacterial pathogens. *Small* 4: 481–484.

253. Kar, S., Bindal, R.C., Tewari, P.K. 2012. Carbon nanotube membranes for desalination and water purification: Challenges and opportunities. *Nano Today* 7: 385–389.

254. Goh, P.S., Ismail, A.F., Ng, B.C. 2013. Carbon nanotubes for desalination: Performance evaluation and current hurdles. *Desalination* 308: 2–14.

255. Goh, P.S., Ismail, A.F. 2015. Graphene-based nanomaterial: The state-of-the-art material for cutting edge desalination technology. *Desalination* 356: 115–128.

256. Mahmoud, K.A., Mansoor, B., Mansour, A., Khraisheh, M. 2015. Functional graphene nanosheets: The next generation membranes for water desalination. *Desalination* 356: 208–225.

257. Ruan, M., Hu, Y., Guo, Z., Dong, R., Palmer, J., Hankinson, J., Berger, C., de Heer, W.A. 2012. Epitaxial graphene on silicon carbide: Introduction to structured grapheme. *MRS Bull.* 37: 1138–1147.

258. Leenaerts, O., Partoens, B., Peeters, F.M. 2008. Graphene: A perfect nanoballoon. *Appl. Phys. Lett.* 93: 193107.

259. Lu, Q., Huang, R. 2009. Nonlinear mechanics of single-atomic layer graphene sheets. *Int. J. Appl. Mech.* 1: 443–467.

260. Xu, Z., Gao, C. 2011. Graphene chiral liquid crystals and macroscopic assembled fibres, *Nat. Commun.* 2: 571.

261. Zaib, Q., Fath, H. 2012. Application of carbon nano-materials in desalination processes, *Desalin. Water Treat.* 51: 627–636.

262. Geim, A.K. 2009. Graphene: Status and prospects, *Science*, 324.

263. Lee, C., Wei, X., Kysar, J.W., Hone, J. 2008. Measurement of the elastic properties and intrinsic strength of monolayer graphene. *Science* 321: 385–388.

264. Bunch, J.S., Verbridge, S.S., Alden, J.S., van der Zande, A.M., Parpia, J.M., Craighead, P.L. McEuen, H.G. 2008. Impermeable atomic membranes from graphene sheets. *Nano Lett.* 8: 2458–2462.

265. Cohen-Tanugi, D., Grossman, J.C. 2012. Water desalination across nanoporous graphene. *Nano Lett.* 12: 3602–3608.

266. Konatham, D., Yu, J., Ho, T.A., Striolo, A. 2013. Simulation insights for graphene-based water desalination membranes, *Langmuir* 29: 11884–11897.

267. Wang, E.N., Karnik, R. 2012. Graphene cleans up water. *Nat. Nanotechnol.* 7: 552–554.

268. Bae, S., Kim, H., Lee, Y., Xu, X., Park, J., Zheng, Y., Balakrishnan, J., Lei, T., Kim, H.R., Song, Y.I., Kim, Y., Kim, K.S., Özyilmaz, B., Ahn, J.-H., Hong, B.H., Iijima, S. 2010. Roll-to-roll production of 30-inch graphene films for transparent electrodes. *Nat. Nanotechnol.* 5: 574–578.

269. Cohen-Tanugi, D., Grossman, J.C. 2012. Water desalination across nanoporous graphene. *Nano Lett.* 12: 3602–3608.

270. Suk, M.E., Aluru, N.R. 2010. Water transport through ultrathin graphene. *J. Phys. Chem. Lett.* 1: 1590–1594.

271. Kannam, S.K., Todd, B.D., Hansen, J.S., Daivis, P.J. 2012. Slip length of water on graphene: Limitations of non-equilibrium molecular dynamics simulations, *J. Chem. Phys.* 136: 024705.

272. Xiong, W., Liu, J.Z., Ma, M., Xu, Z.P., Sheridan, J., Zheng, Q.S. 2011. Strain engineering water transport in graphene nanochannels. *Phys. Rev. E.* 84: 05632.

273. Dhiman, P., Yavari, F., Mi, X., Gullapalli, H., Shi, Y.F., Ajayan, P.M., Koratkar, N. 2011. Harvesting energy from water flow over graphene. *Nano Lett.* 11: 3123–3127.

274. Gordillo, M.C., Marti, J. 2010. Water on graphene surfaces. *J. Phys. Condens. Mater.* 22: 284111.

275. Hu, M., Mi, B. 2013. Enabling graphene oxide nanosheets as water separation membranes, *Environ. Sci. Technol.* 47: 3715–3723.

276. Tang, C.Y., Zhao, Y., Wang, R., Hélix-Nielsen, C., Fane, A.G. 2013. Desalination by biomimetic aquaporin membranes: Review of status and prospects. *Desalination* 308: 34–40.

277. Zhaoa, X., Jianga, Z., Lia, Z., Fana, X., Zhua, J., Wua, H., Yanlei, S., Yanga, D., Pana, F., Shi, J. 2014. Biomimetic and bio-inspired membranes: Preparation and application. *Prog. Polym. Sci.* 39: 1668–1720.

278. Shen, Y-X, Saboe, P.O., Sines, I.T., Erbakan, M., Kumar, M. 2014. Biomimetic membranes: A review. *J. Membr. Sci.* 454: 359–381.

279. Agre, P., Sasaki, S., Chrispeels, M.J. 1993. Aquaporins: A family of water channel proteins. *Am. J. Physiol. Ren. Physiol.* 265: F461.

280. Meinild, A., Klaerke, D., and Zeuthen, T. 1998. Bidirectional water fluxes and specificity for small hydrophilic molecules in aquaporins 0–5. *J. Biol. Chem.* 273: 32446.

281. Agre, P., Preston, G.M., Smith, B.L., Jung, J.S., Raina, S., Moon, C., Guggino, W.B., Nielsen, S. 1993. Aquaporin CHIP: The archetypal molecular water channel. *Am. J. Physiol.Ren. Physiol.* 265: F463–F476.

282. Zhu, F.Q., Tajkhorshid, E., Schulten, K. 2004. Theory and simulation of water permeation in aquaporin-1. *Biophys. J.* 86: 50–57.

283. Kaufman, Y., Berman, A., Freger, V. 2010. Supported lipid bilayer membranes for water purification by reverse osmosis. *Langmuir* 26: 7388–7395.

284. Wang, H., Chung, T-S., Tong, Y.W. 2013. Study on water transport through a mechanically robust AquaporinZ biomimetic membrane. *J. Membr. Sci.* 445: 47–52.

285. Zhonga, P.S., Chunga, T.-S., Jeyaseelanb, K., Armugamb, A. 2012. Aquaporin-embedded biomimetic membranes for nanofiltration. *J. Membr. Sci.* 407– 408: 27–33.

286. Zhao, Y., Qiu, C., Li X., Vararattanavech, A., Shen, W., Torres, J., Hélix-Nielsen, C., Wanga, R., Hua, X., Fane, A.G., Tang, C.Y. 2012. Synthesis of robust and high-performance aquaporin-based biomimetic membranes by interfacial polymerization-membrane preparation and RO performance characterization. *J. Membr. Sci.* 423–424: 422–428.

287. Li, X., Wang, R., Wicaksana, F., Tang, C., Torres, J., Fane, A.G. 2014. Preparation of high performance nanofiltration (NF) membranes incorporated with aquaporin Z. *J. Membr. Sci.* 450: 181–188.

288. Tewari, P.K. 2016. Nanocomposite membrane in water treatment. *Nanocomposite Membrane Technology: Fundamentals and Applications*, ed. P.K. Tewari. CRC Press, Taylor & Francis Group, USA.

5 Desalination

5.1 INTRODUCTION

Desalination is the process that produces pure water from saline water using a certain amount of energy. The minimum thermodynamic energy requirement for desalinating saline water containing 35,000 ppm total dissolved solids (TDS) with zero recovery is about 0.7 kWh/m^3 for a reversible process. Several water-stressed and arid regions are augmenting their water supply with desalinated water to meet the water demand.

Desalination processes can be classified on the basis of phase change and energy used:[1]

Phase change:

- Processes without phase change, such as reverse osmosis (RO) and electrodialysis (ED).
- Processes with phase change, such as multi-stage flash (MSF), multi-effect distillation (MED), vapor compression (VC) and freezing.

Type of energy used:

- Processes using thermal energy, such as MSF, MED and thermal VC.
- Processes using mechanical energy, such as RO and mechanical VC.
- Processes using electrical energy, such as ED.

The actual amount of energy required for separating 1,000 L of fresh water from seawater varies from about 3 kilowatt hour (kWh) to about 16 kWh depending on the type of process and design constraints.

Commercially established desalination processes are mainly thermal and membrane processes.[2] MSF, MED and VC are the commercial thermal desalination processes that utilize heat energy for seawater desalination. RO and ED are the membrane processes for water desalination. RO uses neutral membrane and mechanical energy (pressure) to achieve separation of pure water from saline water. ED uses ionic membranes and electrical energy (potential).

5.2 GLOBAL DESALINATION SCENARIO

The worldwide research effort has been led by the US government since the early 1960s through the creation and funding of the Office of Saline Water and its successor organizations. By the late 1960s, small- and medium-capacity commercial thermal approaches to desalting water had been established. In the 1970s, commercial-scale membrane processes such as reverse osmosis (RO) and electrodialysis (ED) were

TABLE 5.1

Water quality guidelines for human consumption with respect to salinity

Use	Rating	Salinity Range (mg/L TDS)
Human consumption	Excellent	About 100
Human consumption	Good to fair	100–1,000
Human consumption	Poor	1,000–1,200
Human consumption	Unacceptable	More than 1,200
Irrigation depending on plant	Maximum limit for healthy growth	3,500

introduced and used extensively. As technology progressed and operational experience increased, construction and operation costs reduced significantly. As the scarcity and price of conventional sources of fresh water increased with time, desalination has gathered pace as an option for producing fresh water for human consumption, industries and other uses.

Desalination plants have the potential to produce and supply good-quality water for drinking and water of appropriate quality for industry. With advances in desalination technology, more and more industries are coming forward to invest in desalination plant to meet their water requirements. Table 5.1 gives the water quality guidelines with respect to salinity for different end uses.

The global cumulative contracted desalination capacity in 1980 was about 7,000 million liters per day (MLD), which had increased by several times to 95,000 MLD in 2016. Global online desalination capacity was about 5,000 MLD in 1980, increasing to 88,000 MLD by 2016, implying that the growth of desalination is quite substantial.[3] It is estimated that almost 85% of total desalination capacity is contributed by sea water desalination. The remainder is brackish water desalination. The contribution of oil-rich Arab states to global desalination capacity is quite high. It was about 70% in 1985, 50% in 1995 and 40% in 2005, indicating that desalination processes are playing an increasingly important role in augmentation of water supply in other parts of the world. The growth of desalination in Asia and southern Europe is quite appreciable. MSF, MED and RO are the major commercial processes contributing to overall capacity. Membrane processes (particularly RO) account for about 60% of the installed capacity, while thermal processes account for about 40%. However, the unit capacities of thermal processes are high compared to those of membrane processes.

The increase in desalination capacity is mainly due to increased water demand and a significant reduction in the cost of desalination due to technological advances which have made desalinated water cost competitive. In some places, desalination successfully competes with conventional water resources and water transport for potable water supply. Thermal and membrane-based desalination technologies are extensively used, depending on the site conditions and user requirements, and statistics indicate remarkable development levels in both processes.

The cost of desalination is site specific and depends on several parameters such as feed water salinity and other water-quality parameters, plant capacity, energy cost, labor cost, type of contract, political and environmental restrictions, and other local factors.[4] Although the cost of desalinated water is normally higher than municipal or conventional water supplies, its reliability and sustained quality has made it the imperative choice for critical requirements. The appropriate technology for any particular location should be chosen using a uniform evaluation methodology.[5] MSF, MED and RO are the established commercial desalination processes significantly contributing to overall capacity.[6] Thermal technologies such as MSF and MED are preferable where energy is available at low cost, and raw water is polluted or of high salinity. Thermal technologies have a proven record of reliability, dependability and potential for cogeneration of power and water. However, several recent RO plants have been installed along with new power plants rather than cogeneration.

5.3   COMMERCIAL DESALINATION PROCESSES

MSF, MED and RO are the main desalination processes currently dominating the commercial desalination market.

5.3.1   Multi-Stage Flash

Multi-stage flash (MSF) is an established thermal desalination process with proven performance and robust operation. The standard design is available commercially. MSF is a preferred technology for large-capacity installations where thermal energy is available.

MSF is based on the principle of evaporation and condensation. Sea water is heated to a particular temperature using the condensation heat of the vapors produced, supplemented with external steam as the heat source. The heated sea water is flashed in successive stages maintained at relatively lower pressure. The vapor produced is condensed and recovered as product water.

In a typical MSF unit, sea water is screened for the removal of suspended solids and dosed with chlorine to maintain a level of about 0.2 ppm free chlorine at the inlet of the MSF evaporator. Chemical pretreatment in the form of either a chemical additive or acid treatment suppresses the formation of alkaline scale in the heat transfer tubes. It is deaerated to remove dissolved oxygen and carbon dioxide to minimize corrosion and improve heat transfer performance. After pretreatment, sea water flows through the tubes of the MSF evaporator and the condensation heat from water vapor is utilized for preheating. It is heated up to the designed temperature using low-pressure steam in the heat input section, known as the brine heater. It then enters into the first stage of the flash evaporator. The top-brine temperature ranges between 90°C and 120°C depending on the feed pretreatment method. The advantage of higher top-brine temperature is high productivity, i.e., production of a higher amount of desalinated water per unit of steam consumption or a high gain output ratio (GOR), and hence better steam economy.

An MSF evaporator is divided into several compartments called flash stages. These stages are maintained at successively reduced pressures. When the brine enters into a flash stage maintained below the saturation vapor pressure of water, a fraction of it flashes into vapor. The vapor passes through the mist eliminator and condenses on the outer surface of the heat transfer tubes, giving its latent heat to incoming sea water flowing inside the tubes. The unflashed brine moves to the next flash stage, and the process is repeated. The desired vacuum is maintained with the help of a steam jet ejector. The brine keeps flashing at every subsequent stage up to the last stage due to successively reducing pressure. From the last stage, the concentrated brine is blown down through a pump. The water vapor produced in each flash stage is cooled and condensed by seawater flowing through the tubes. The condensate collected in each stage eventually flows to the last stage and is drawn as product water.

Copper-based alloys such as cupronickel and titanium are normally used as heat transfer material for the tubes and tubesheets in the brine heater and MSF evaporator. The MSF evaporator and brine heater shell are made of stainless steel, cupronickel or carbon steel with lining/cladding. Centrifugal pumps made of stainless steel (AISI316) or similar corrosion-resistant material are used for pumping the brine. Two major flow arrangements are used in MSF: the once-through and the brine recycle flow. The majority of operational MSF systems are based on brine recycling. In the recycle type, the make-up feed is mixed with the recirculation brine and preheated in the heat recovery section by recovering the energy of the condensing vapor. A heat reject section at the cooling end of the plant removes the quantity of waste heat required for steady operation. The once-through system requires more chemicals and feed sea water than the recycle system. The MSF desalination system is illustrated in Figure 5.1.

Sea water contains sulfate, bicarbonate, calcium and magnesium ions, which may contribute to scaling. On heating, the bicarbonate ions decompose to carbonate ions, resulting in alkaline ($CaCO_3$ and $Mg(OH)_2$) scale formation. Formation of scale leads to deterioration in the overall heat transfer performance and production rate. Hence, the feed sea water is pretreated with acid or chemical additives to control alkaline scale formation. In the case of acid treatment, the bicarbonate ions are decomposed to carbon dioxide. In chemical additive dosing, certain proprietary compounds are added to the seawater feed, changing the scale characteristics so that loose, easily removable scale is formed. Calcium sulfate ($CaSO_4$) scale formation,

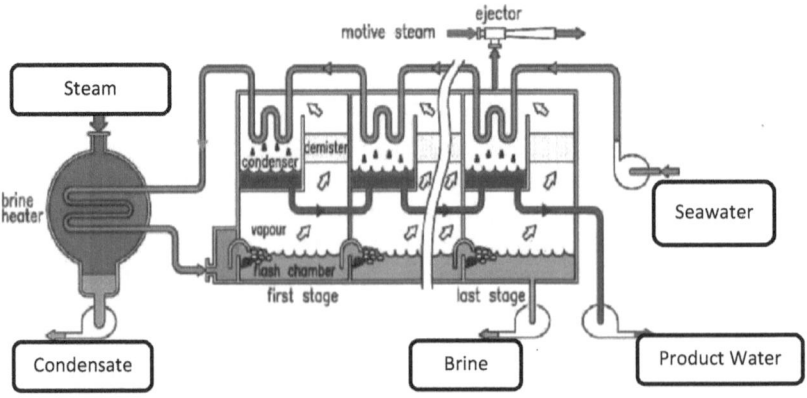

FIGURE 5.1 Multi-stage flash desalination system.

which is also sometimes referred to as hard scale, is to be avoided in desalination plants. It is insoluble in mineral acids and difficult to remove by acid cleaning. Only mechanical methods or expensive chemical treatment can be used for its removal. The allowable top-brine temperature and concentration is controlled to avoid $CaSO_4$ scaling. The top-brine temperature in additive-based plants is maintained between 90°C and 110°C, whereas in acid-type plants it is limited to 120°C. The temperature rise across the brine heater is fixed on the basis of cost optimization with respect to the GOR and the number of flash stages. The GOR is the ratio of desalinated water produced per unit of steam consumed. The highest GOR value that can be achieved in MSF is around 12, because of loss in vapor temperature due to boiling point rise, pressure loss through demisters, condenser tube bundle and non-equilibration losses. Two types of tube bundle configurations are used in commercial MSF evaporators. The majority of them use a cross-tube arrangement, where tubes are laid at right angles to the flow direction of the flashing brine. In long-tube configurations, tubes are parallel to the flow direction of the flashing brine. MSF heat transfer area requirements are usually in the range of 20–140 $m^2/(m^3/h)$ desalted water produced, depending on GOR. The overall heat transfer coefficient varies from 2.2 to 5 $kW \cdot m^{-2} \cdot K^{-1}$. MSF plants produce practically pure distilled water with 1–5 ppm total dissolved solids (TDS) from sea water containing 35,000–45,000 ppm TDS.

MSF is known as the work-horse of the desalination industry. The process can accept higher contaminant loading of suspended solids, heavy metals, oil and grease, COD, BOD, etc. in seawater feed. It produces distilled-quality product water of high purity that is good for high-end applications in power plants and process industries. Future development in seawater desalination is directed towards reducing overall cost. Innovation in the following areas is required:

(i)   reduce energy consumption,
(ii) utilize low-grade heat as energy input for sea water desalination,
(iii) develop improved, cheaper, corrosion-resistant material,
(iv) pretreatment of the sea water by membrane processes such as nanofiltration or antiscalants to enable plants to operate at higher top-brine temperature,
(v)   scale-up.

The MSF process, despite several improvements, is characterized by relatively lower performance ratio or GOR (typically 8) than multiple-effect evaporation (typically 12). Long-tube design and higher temperature operation have been explored for performance enhancement. Operation of MSF at higher temperature is often limited by scaling problems. Using a nanofiltration pretreatment to remove scale-causing compounds has potential to minimize scaling caused by deposition of calcium sulfate. Corrosion at higher temperatures is also a matter for concern, as high-quality material resistant to pitting and crevice corrosion from aggressive seawater is required.

Comprehensive and advanced mathematical models may be developed for process analysis and subsequent improvement in process efficiency, taking into account the physico-chemical aspects of the process including stage geometry, variability of saline water characteristics, heat transfer aspects, fouling and process non-idealities etc.

5.3.2 MULTI-EFFECT DISTILLATION

Multi-effect distillation (MED) has a long history. It was the first thermal desalination process to be used to produce significant amounts of fresh water from sea water. However, it is only in recent decades that large-scale application has begun. In an MED unit (Figure 5.2), each effect operates at a successively lower temperature and pressure.

In a low-temperature MED unit, the first effect is heated by low-pressure steam, typically at about 0.3 bar. The heat source can be turbine exhaust steam or extraction steam. In the first effect, vapor is generated from the feed water. The vapor is directed to the second effect. Vapor from the previous effect serves as a heat source for the succeeding effect to evaporate the brine. Vapor from the last effect is condensed as product water in the final condenser where sea water is used as coolant. The vapor produced in each effect passes through demisters and condenses inside the tubes as product water, transferring the latent heat to the brine falling outside the tubes, enabling a portion of the brine to evaporate. A low-temperature MED unit operates at about 55°C–65°C, enabling cheaper construction material, as there will be fewer corrosion and scaling problems than with high-temperature thermal desalination plants.

MED plants can be built in vertical-tube or horizontal-tube arrangements. In the vertical-tube evaporator (VTE), the sea water boils in a thin film flowing inside the tube and steam condenses on the outer side of the heat transfer tube. A VTE may consist of falling film evaporator, rising film evaporator and evaporator with forced and natural circulation. A significant enhancement in heat transfer performance may be achieved by using fluted tubes to extend the surface and reduce the film thickness. Normally, multiple effects are used in series in a desalination plant.

Most current MED plants have horizontal-tube evaporator units. The horizontal-tube thin-film (HTTF)-type MED is quite popular commercially. In the HTTF evaporator, the seawater feed is sprayed on the outer surface of the tubes and vapor flows inside the horizontal tubes, where it condenses into product water. The advantage is that the overall heat transfer coefficient is about three times that of a submerged-tube desalination plant. However, HTTF evaporation is sensitive to the distribution of brine on tubes, and the seawater supply to each effect should be sufficient to prevent dry spots.

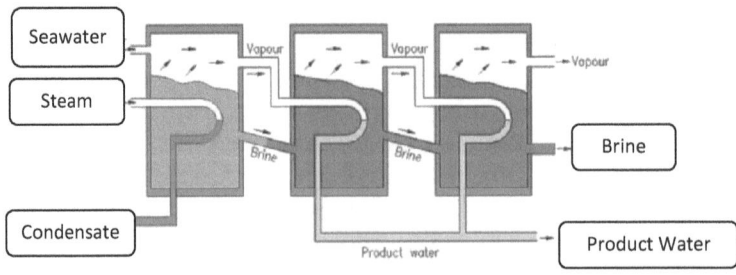

FIGURE 5.2 MED process flow diagram.

MED plants using polymer additive-type scale control are generally designed for low-temperature operation. The performance of the low-temperature MED plant can be improved still further by vapor compression. In large plants, thermo-compression is used to compress the vapor. Higher-pressure steam is used as a motive fluid in the thermo-compressor to compress part of the vapor from the last effect, which is used as a heating medium in the first effect. MED construction materials are generally similar to those for MSF. Some suppliers have standardized on aluminum alloy heat transfer tubes and epoxy-coated carbon steel plates for the shell.

The MED plant produces high-purity distilled-quality water, similar to MSF. The low-temperature operation, waste heat utilization, low cooling water requirements and low energy consumption make MED an attractive alternative for sea water desalination. The low-pressure extraction steam at 0.35 bar pressure from power plant is sufficient to desalinate seawater using the MED process.

HTTF evaporators, however, can operate at higher temperatures (up to 130°C), using more expensive materials. A high-temperature MED (HT-MED) can incorporate a larger number of effects and, therefore, have a greater GOR. MED plants usually have a GOR in the range of 6–12. HT-MED plants may have GOR of 20 or higher. Selection of the GOR for MED is a matter of economic optimization. The MED unit can be made more energy efficient by increasing the number of effects and the heat transfer area, or by increasing the operating brine temperature. On the other hand, it is often the case that lower water costs are achieved by operating at lower temperatures (LT-MED), because of the resultant decrease in corrosion and scaling, which permits the use of cheaper construction materials.

The MED process produces high-quality (around 1–5 ppm TDS) product water from 35,000–45,000 ppm TDS of seawater. The energy efficiency of the MED unit increases as the number of effects increases. The GOR for MED is theoretically equal to the number of effects, but in practice, it is somewhat less due to the heat losses. Like MSF, the GOR for MED varies only slightly with the salinity of the feed. Enhanced heat transfer coefficient, enhanced boiling heat transfer at low wall superheat and development of materials to minimize corrosion and increase life are needed.

MED has been used in industrial applications for several years at many locations worldwide. Its major advantage is the ability to produce a significantly higher GOR than MSF. It has the potential to be one of the dominant processes for thermal desalination in the small- and medium-capacity ranges.

5.3.3 Vapor Compression

Vapor compression (VC) is an energy-efficient thermal process for sea water desalination. Depending on the type of motive energy used for vapor compression, there are two variants of this process: mechanical vapor compression (MVC) and thermo-compression (TC).

In a typical VC unit, the vapor produced in the desalination evaporator is compressed and heated by the vapor compressor. Seawater feed is pumped to the low-pressure side of the evaporator, where it picks up heat and boils. The compressor heats the water vapor by the heat of compression and delivers it to the high-pressure side of the evaporator, where it condenses by giving up its latent heat to the boiling sea water. The efficiency is governed by the effectiveness of the heat transfer surface, the compressor efficiency and the effectiveness of heat recovery from the reject brine and product water streams.

Heat for evaporating the brine comes from vapor compression rather than directly from externally supplied steam. The VC units are built in a variety of configurations. In a TC unit, high-pressure steam is used as the motive fluid for the compression of vapor in an ejector. High-temperature saturated vapor is used to evaporate a part of the brine in the evaporator. The condensate is pumped out as the product water. VC desalination units can be designed to achieve high GOR. The units are normally designed to operate at top-brine temperature of up to 65°C temperature and 0.3 bar pressure. It produces high-quality product water for high-end applications.

MVC desalination units are characterized by low raw water requirement and high efficiency. However, due to the restrictions on the compressor size, large-capacity applications for sea water desalination are limited and plants up to 3 MLD capacity have been built. Heat transfer rates can be increased by using mechanically spinning heat-exchange surfaces. Efforts are being directed towards the development of high-efficiency compressors.

MED plants give high GOR, especially in combination with vapor compression. Some of the important areas of considerations in MED are:

- Low-temperature operation of MED is quite promising, despite the advantages of a high-temperature multiple-effect desalination (HT-MED) plant. Operation at low temperature avoids scaling and corrosion problems. The low-operating top-brine temperature (TBT) results in a desirable outcome such as low energy consumption (as low as 2.5 kWh/m^3) when using low-grade or waste heat and up to 5 kWh/m^3 when using prime energy. Since a low temperature plant can make effective use of low-cost, low-grade heat or even zero-cost waste heat, the energy cost is reduced.
- Existing MED units combined with thermal vapor compression (TVC) has potential to give a high GOR ranging from 12 to 16. Field data for MED combined with mechanical vapor compression (MVC) show competitive (at times favorable) power consumption. The main drawbacks of the MVC are limitations on the compressor range, which limits operation to low and medium capacities.
- Development of efficient tubes has the potential to enhance the heat transfer coefficient and heat transfer performance of the plant.

5.3.4 Reverse Osmosis

Reverse osmosis (RO) is a membrane-based desalination technology operating at ambient temperature, requiring electrical power for the high-pressure pump. In the RO process, pure water permeates from saline water through a semipermeable membrane under pressure (Figure 5.3).

Osmosis is a natural phenomenon by which water flows through a semipermeable membrane from a dilute to a concentrated solution. Osmotic equilibrium is

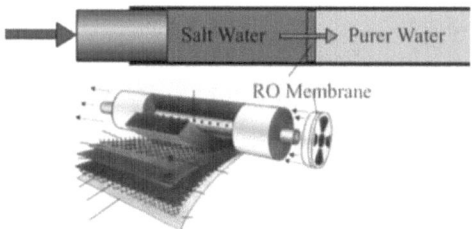

FIGURE 5.3   Reverse osmosis process.

achieved when the flow due to osmosis is just stopped by the application of pressure to the high solute concentration side. The corresponding pressure is called the osmotic pressure. When the pressure of saline water is higher than the osmotic pressure, pure water passes through the semipermeable membrane, leaving the concentrate stream behind. RO is a pressure-driven membrane process using a semipermeable membrane that preferentially permeates solvent. The mechanism of water permeation through the semipermeable membrane in RO is related to the membrane's chemical nature and morphology. As a rule of thumb, the osmotic pressure of the sodium chloride solution (expressed in psi) is 1% of the salt concentration (expressed in ppm).

A typical RO desalination plant consists of a pretreatment system, a high-pressure pump, RO modules and a post-treatment system. On the basis of feedwater analysis, a suitable pretreatment system is designed and used to take care of suspended solids, biofouling and scaling possibilities. Fouling and scaling adversely affect the performance of the membrane and shorten its life. The silt density index (SDI) is a measure of the fouling constituents in the seawater feed such as iron, manganese, organics and fine suspension. Feedwater with an SDI value of 2–3 is preferred; a value of 4 is on the high side, and 5 produces an undesirable level of fouling and scaling on the membrane surface.

Sea water is passed through a conventional filter system consisting of clarifier, sand filter, activated carbon filter and cartridge filters, before being pressurized by the high-pressure pump (see Figure 5.4). Coagulant aids and polyelectrolytes are added in the clarifier to remove suspended solids and colloids. Acid, antiscalant and sodium bisulfite are added to achieve higher water recovery and to maintain membrane performance. The pressurized water passes through a series of membrane elements, packed in pressure vessels called modules. Pure water permeating through the membrane elements is sent to the post-treatment system, depending on the end use. The reject water, which constitutes about 50%–65% of the feed, is about 1.5 bar less than the pressurized feed water. The energy content of this stream can be recovered using an energy recovery system. The module arrangement, operating pressure and the percent recovery are governed by optimal process design parameters. RO normally uses only electrical energy, and the largest power consumer is the high-pressure pump, which delivers flow at a pressure of 50–80 bar. In large-capacity RO desalination plants, it is possible to recover a significant amount of the energy from high-pressure reject brine using an energy recovery system such

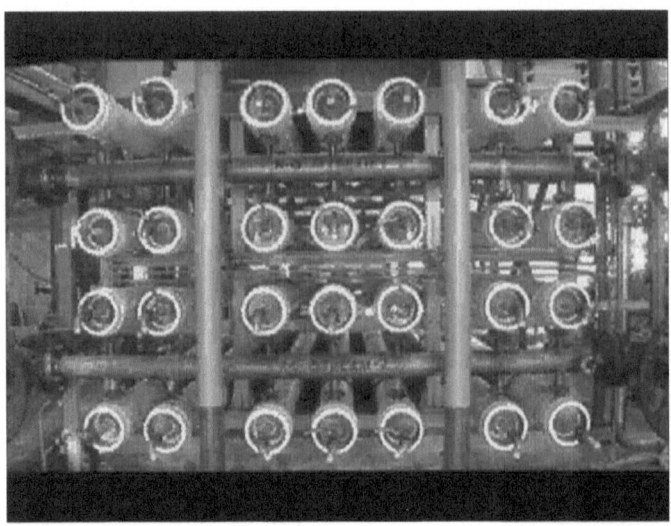

FIGURE 5.4 Reverse osmosis desalination system.

as pressure exchanger, pelton wheel, hydro turbine or turbocharger. The energy consumption in a seawater RO plant using an energy recovery unit is around 2.5–6 kW(e)·h/m³ of product water.

The RO membrane system comprises three fluid streams: feed, permeate and concentrate. As the feed water passes through the membrane, permeate (product water) and concentrate (reject water) are produced. The combined flow rate of permeate (Q_p) and concentrate (Q_c) is equal to the feed flow rate (Q_f).

$$Q_f = Q_p + Q_c$$

Overall mass balance is:

$$Q_f C_f = Q_p C_p + Q_c C_c$$

where C_f is the concentration of salt in feed water, C_p is the concentration of salt in product water and C_c is concentration of salt in concentrate stream.

RO performance is evaluated in terms of the percent recovery, the permeate flux and the percent salt rejection. The recovery is defined as the ratio of permeate flow rate to feed flow rate. The permeate flux is the flow rate of permeate water across a unit area of membrane. The salt rejection describes the quantity of salt removed from the feed water stream. The percent salt rejection (SR) is given by:

$$\%SR = \left(\left(C_f - C_p \right) / C_f \right) \times 100$$

Unlike thermal processes, the quality of raw water has a significant effect on the performance of the reverse osmosis unit. Pretreatment of raw water is an important part of the RO system. It is critical for ensuring reliability and cost-effectiveness.

Current research is looking at advanced systems for pretreatment, higher recovery and salt rejection, enhanced membrane life and chemical resistance, minimizing energy consumption and selecting appropriate construction materials. Hydraulic permeability, selectivity, structural configuration and base polymer chemical characteristics of the membrane are the main parameters determining RO process efficiency.

The following aspects of the RO system require further R&D:

- Fouling mitigation, enabling higher membrane flux due to minimal scaling and fouling of membranes.
- Design flexibility to achieve efficient operation despite seasonal fluctuations in raw water turbidity, particle size distribution, total and organic suspended matter and biological load.
- Addressing scaling issues involving sparingly soluble salts, silica etc.
- Conditions maximizing boron rejection and membranes with improved boron rejection capability.
- Corrosion and corrosion inhibition studies in raw water intake, membrane system, product delivery pipelines and machinery to suggest suitable alternative material/ modified operating conditions.
- Studies on performance of new anti-scaling and anti-foaming agents.
- Membrane failure diagnosis, autopsy and corrective measures to enhance life.
- Brine management and studies on brine discharge.
- Advanced energy recovery devices.

Surface modification by ultraviolet irradiation may result in more hydrophilic membranes with lower fouling tendencies and higher permeability; the resulting lower-pressure operation can potentially reduce both investment and operating costs. Ultrafiltration (UF) and nanofiltration (NF) pretreatment options are being pursued as integrated membrane operations. Exploration continues of UF/NF pretreatment options balancing ion rejection (SO_4^{2-}, HCO_3^-, Mg^{2+}, Ca^{2+}) against TDS reduction and permeate flow rate.

Most commercial desalination processes normally produce two streams: (i) the product water stream, and (ii) the concentrated brine stream. Disposal of the concentrated stream poses environmental issues at times. Brine concentrate either from membrane-based plant or thermal desalination plant is normally free from biological constituents because the feed is pretreated in the case of a membrane-based plant, whereas it is heated to higher temperatures in the case of a thermal desalination plant. It contains higher concentrations of valuable species compared to normal seawater. Hence, it is a good source of valuable materials. Salt manufacturers can use the concentrate after the recovery of valuables, as they would benefit from the higher concentration and at times, higher temperature.

A number of small and large reverse osmosis plants are in operation throughout the world. It is important to develop better membranes offering higher output while maintaining the salt rejection, chemical pretreatment, membrane life and energy requirement. The operating pressure of seawater RO desalination plants has come down significantly from the more than 70 bar that applied in the initial stages of the technology. Present efforts are directed towards improving pretreatment methodology to minimize fouling/ scaling and hence to enhance membrane performance and life. Use of membrane-based microfiltration (MF) and UF processes are being assessed for their ability to reduce biofouling.

The basic cost elements of the RO technology contributing to overall water cost are:

- Energy requirement
- Membrane replacement
- Pretreatment chemicals
- Consumables such as cartridge filter
- Manpower
- Depreciation and interest.

Energy and membrane replacement are the major cost components and may vary from 50% to 60% of the total water cost depending on the configuration, unit capacity and site-specific local factors. Recent developments in energy recovery devices have brought down energy consumption but there is further scope for improvement. Power consumption for the intake point to the pretreatment section is site specific, hence selection of the intake source is an important aspect. Fine-tuning the process energy requirements, including pretreatment, and improving recovery are thus the areas that currently require attention. Membrane-based UF as a pretreatment system has the potential to reduce energy consumption by reducing the number of unit operations.

Unlike thermal processes, most of these plants require operator intervention to maintain proper process parameters and controls due to the variation in feedwater quality. Preserving membrane elements and equipment during shutdown is also an important aspect of RO operation. A built-in mechanism for membrane preservation and equipment safeguarding would reduce maintenance costs and increase the operational life of the plant. So far successful RO membranes have been polyamide based. Polyamide membranes are susceptible to oxidizing agents and fouling. There is a need to develop non-fouling membranes. Surface modification using ultraviolet irradiation appears to have yielded some encouraging results.

The disposal of spent polyamide membranes could be an environmental issue, as it can lead to emission of nitrogen oxides (NO_x). Thus, there is a need to look for fouling-resistant membranes based on eco-friendly polymers.

Co-location of desalination plants alongside power plant or process industries utilizing sea water enables sharing of resources at marginal costs.

5.3.5 ELECTRODIALYSIS

Electrodialysis uses membranes that are selectively permeable to ions, based on their charge. Two types of selective membranes are used, those for cation exchange and anion exchange. Membranes that permeate cations are called cation-exchange membranes. Membranes that permeate anions are known as anion-exchange membranes. These membranes are placed alternately, with flow channels between them. Electrodes

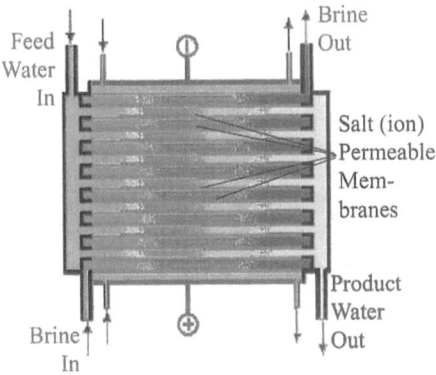

FIGURE 5.5 Electrodialysis flow diagram.

are placed on each side. An electric current is applied to the ED unit. Due to the electric field, the respective membranes allow the cations and anions to pass through. The electrodes draw their counter-ions through the membranes, so that these are removed from the water, thus giving purified water.

In other words, ED is a membrane process in which application of an applied electrical potential selectively removes the dissolved ionic solutes from saline water through charged membranes. A typical ED unit (Figure 5.5) consists of a series of anion- and cation-exchange membranes arranged in an alternating pattern creating individual cell pairs between an anode and a cathode. An ionic solution such as saline water is pumped through the cell pairs and an electrical potential is established between the anode and the cathode. Positively charged cations migrate towards the cathode and negatively charged anions towards the anode. The cations pass easily through negatively charged cation-exchange membranes but are retained by positively charged anion-exchange membranes. Likewise anions pass easily through positively charged anion-exchange membranes but are retained by negatively charged cation-exchange membranes. The overall result is an increase in the concentration of the alternate compartments, while the other compartments become simultaneously depleted. The depleted solution is purified water. The mode of operation is unidirectional, requiring a direct current power supply.

The main components of the ED system are the stacks with electrodes, membranes, power supply unit and feed/product-circulating pumps. The membranes are prepared by dispersing ion-exchange resins in suitable binders. Fabric supports are used to mechanically strengthen the membranes. The spacers are used to provide turbulence and minimize concentration polarization. Effective chemical pretreatment of the feed is required for longer membrane life.

The power consumption is directly proportional to the dissolved ionic solute concentration. ED is widely used for desalination of 1,000–5,000 ppm TDS brackish water. High-temperature ED is also being explored. Higher feed water temperature is useful for seawater desalination because both membrane and solution resistances are reduced and higher current densities are achieved. For sea water containing about 35,000 ppm total dissolved solids, the approximate power required to produce

1,000 L of product water containing about 500 ppm dissolved solutes is about 16 kWh. Electrodialysis for seawater desalination is currently an energy-intensive technology. Only a few small ED plants are in operation for seawater desalination. A few large-capacity ED plants were set up in Japan for concentrating and producing salt rather than desalination. ED is yet to prove itself the system of choice for large-scale sea water desalination due to high energy consumption and the constraint of removing non-ionic solutes from the raw water. With innovations like frequent polarity reversal of the electrodes, popularly known as electrodialysis reversal (EDR), the energy consumption has been brought down significantly but it is still on the high side. It has several advantages over unidirectional ED. It provides a flushing operation for the deposits and controls membrane scaling and fouling. Development of advanced ion-exchange membranes using nanometer-scale self-assembly techniques has the potential to offer higher efficiency and lower electrical energy consumption.

5.3.6 COMPARISON

Some of the key data for commercial desalination processes are given in Table 5.2.

The energy cost represents a significant fraction of the cost of desalted water. Reducing the energy cost is therefore an important goal of any desalination technology.

TABLE 5.2
Key Data for Commercial Desalination Processes

	MSF	MED	MED-TVC	MVC	RO	ED
Top-brine (sea water) temperature (°C)	90–120	55–70	55–70	55–70	Ambient	Ambient
Form of main energy	Steam (heat)	Steam (heat)	Steam (heat)	Mechanical/ electrical energy	Mechanical/ electrical energy	Electrical energy
Thermal energy consumption (kWh/m³)	12	6	10	Not applicable	Not applicable	Not applicable
Electrical energy consumption (kWh/m³)	3.5	1.5	1.5	8–14	1–7	1–20
Typical salt content of raw water (ppm TDS)	30,000– 50,000	30,000– 50,000	30,000– 50,000	30,000– 50,000	1,000– 40,000	1,000– 40,000
Product water quality (ppm TDS)	1–10	1–10	1–10	1–10	200–500	200–500
Current, typical Single-train capacity (m³/d)	5,000– 55,000	500–12,000	100–12,000	10–2,500	1–10,000	1–1,000

Waste heat from nuclear research reactors and nuclear power plants is a good source of low-cost energy for desalination technologies, which can operate effectively at low temperatures. Low-grade heat and waste heat can be used as energy input for sea water desalination. Ocean thermal energy can also be utilized for sea water desalination. Work on long-tube evaporators, leakproof concrete shells and plastic heat exchangers for MED, and long-tube design for MSF is being pursued to reduce the cost of desalinated water. Basic studies show that membrane distillation using thermal driving force has potential for seawater desalination and brine management.

5.4   NUCLEAR DESALINATION

Nuclear desalination[7] means production of potable water from sea water in a complex where a nuclear reactor is used as a source of energy for the desalination process. Electrical and/or thermal energy is used in the desalination process. The facility is dedicated to the production of potable water or used for the generation of electricity and the production of potable water. In either case, it is an integrated facility and energy is produced onsite for use in the desalination system. It involves some degree of common or shared facilities, services, staff, operating strategies, outage planning and control facilities, and seawater intake and outfall structures.

Interest in using nuclear energy for seawater desalination is growing worldwide. This is motivated by the economic competitiveness of nuclear energy, the conservation of limited fossil-fuel resources for environmental protection, carbon footprint reduction and the spin-off effect of nuclear technology in industrial development. Nuclear desalination has great potential to play an important role in ensuring water and energy security.

Nuclear desalination involves three technologies: nuclear reactor technology, desalination technology and coupling aspects. Normally, any type of nuclear reactor can be used for nuclear desalination. Some reactor types, like water-cooled reactors, may be preferred for this application due to their advanced state of development and deployment. These reactors are well proven and the fuel, such as natural uranium or low-enriched uranium, is widely available. Desalination technologies such as MSF, MED and RO have their own strengths. MSF is the preferred technology for large capacity; however, RO is increasing its market share day by day. There is likely to be enhanced use of MED and hybrid systems. The coupling aspects and method of coupling have a significant technoeconomic impact on the overall system, particularly in the case of thermal coupling, when a nuclear reactor is used to supply steam for seawater desalination. It must be ensured that any variation in steam demand from the desalination system would not cause a hazardous situation in the nuclear system. Adequate safety precautions must be taken to prevent the release of harmful radioactivity to the product water.

A cogeneration plant, producing both electricity and water, has several economic advantages over single-purpose plants. Its inherent design strategies optimize both thermodynamic and economic efficiency. The specific financial investment is lower owing to the sharing of facilities. Specific fuel and manpower costs are lower because total costs are divided between the two products. However, a cogeneration plant has the disadvantage of less flexibility and lower availability.

The selection of desalination technology depends on the nuclear reactor type and the quality of product water required by the user, such as drinking, industrial or commercial use. Thermal desalination systems (MSF, MED) provide distilled-quality water that is directly usable for industrial process applications. RO systems provide potable-quality water. If the cogeneration option is chosen, careful consideration should be given to the power-to-water ratio, which is defined as power required per m^3/d of fresh water produced. This parameter may be intrinsically limited by the reactor plant design and could thus affect reactor plant selection. The required power-to-water ratio varies from country to country and also seasonally. The overall safety of an integrated nuclear desalination plant complex is primarily dependent on the safety of the nuclear reactor, the design impact of integrating the reactor plant with the desalination plant and the transient interactions. A particular concern for nuclear desalination plants is the potential for release of radioactive materials into the product water. This can be effectively overcome by designing multiple barriers and pressure differentials to block the transport of radioactive materials to the desalination plant.

Radioactivity in the intermediate system and desalination plant is regularly monitored. In general, proximity to population centers is an advantage for desalination plants from the point of view of water distribution, whereas isolation from population centers is preferable for nuclear plant from the safety point of view. Balancing the two competing factors is an essential element in the overall water and energy supply policy. If the nuclear desalination plant is located near a population center, this must be taken into account in civil emergency preparedness planning. Licensing of nuclear desalination plants requires authorities to deal with water supply and quality regulations with respect to desalination in addition to the body of nuclear regulations that applies to conventional nuclear plants. Regulatory authorities must also address the environmental impacts of nuclear desalination systems, because desalination systems discharge brine and some chemicals in addition to those from the nuclear plant. Good experience exists on the management of these discharges. Nevertheless, they should be carefully assessed to ensure that no significant synergistic impacts can occur.

Experience with nuclear desalination has exceeded 100 reactor-years. There are nuclear desalination plants in India, Kazakhstan and Japan, among others. No safety incidents associated with nuclear desalination have been reported. There is a need to intensify development efforts in the field of nuclear desalination, particularly in the following areas:

(a) Increasing energy efficiency
(b) Optimizing coupling and integration aspects
(c) Development of high-performance coupling systems
(d) Development of small and medium-sized nuclear reactors
(e) Development of next-generation desalination systems
(f) Low-grade and waste heat utilization
(g) Hybrid desalination technologies
(h) Recovery of uranium and valuable materials from brine and seawater
(i) Environmental aspects

The outcome of these development efforts will help to (i) bring down the carbon footprint; (ii) make water technologies more affordable; (iii) address environmental concerns; and (iv) cogenerate water and electricity. Climate-change mitigation policies currently adopted by several countries have still further enhanced interest in nuclear cogeneration projects. However, more effort is required if the deployment of such projects is to effectively penetrate the heat and transportation market, which will require involvement of stakeholders, and appropriate business models and relationships between vendors and users (i.e., the nuclear utility, vendors for nuclear and applications). Better understanding of the requirements and constraints of users and vendors, among other stakeholders involved in nuclear cogeneration projects, is important for facilitating their effective implementation. Such requirements are related to, but not limited to, siting, target utilization of the cogenerated commodity, status and selection of the technologies for the coupled industrial plant, and financial constraints and considerations.

5.5   LOW-CARBON DESALINATION

The energy consumed by desalination systems is about 0.2% of global energy consumption. It is estimated that current online desalination capacity consumes about 200 TWhe/year, or an average power demand of around 23 GWe, and preliminary estimates show a direct carbon footprint of about 120 million metric tons annually. About 41% of this energy is consumed as electricity; the remainder is heat used to drive thermal desalination plants, typically in the form of steam at temperatures between 65°C and 130°C, depending on the technology involved. With RO, about 2.1–3.6 kg CO_2 is produced per 1,000 L of fresh water, depending on the type of fossil fuels used to produce electricity. Thermal desalination technologies such as MED-TC or MSF generally emit 8–20 kg CO_2/m^3 of fresh water (Table 5.3).

The energy required for desalination varies depending on the technology used as well as the salinity and characteristics of the raw water being desalinated. The carbon footprint of a desalination plant depends upon its efficiency and the source of energy that drives it. As in many industries, desalination plants produce indirect greenhouse gas (GHG) emissions as well. The opportunity exists to guide future developments in ways that minimize energy use and greenhouse gas emissions.[9]

If a desalination system is more energy efficient, it implies that a small power input is needed. This means reduced capital cost for energy supply and a lower cost of desalinated water. Renewable sources available at large scale and affordable cost

TABLE 5.3
Representative Direct Greenhouse Gas Footprint[8]

Process	kg CO_2 per 1,000 L of fresh water
Reverse osmosis (RO)	2.1–3.6
Multi-effect distillation thermo-compression (MED-TC)	8–16
Multi-stage flash (MSF)	10–20

include wind power, photovoltaic (PV) power and concentrating solar power (CSP). For wind and PV power, battery storage remains costly, whereas for CSP, thermal energy can be stored relatively inexpensively. While water storage is relatively inexpensive, intermittent use of a desalination plant to meet baseload water demand requires over-sizing the plant. The potential of solar PV and wind power-based desalination systems with a near-zero carbon footprint appear quite promising. Electrical energy storage for renewables needs further development from a techno-economics point of view.

Nuclear energy is the main proven non-renewable low-carbon technology for large-scale baseload power generation. It produces electricity at a reasonably competitive price. However, in some countries it is met with widespread concern, and faces political and social challenges.

Enhanced geothermal energy has the potential to be useful for thermal desalination in some areas. A wide range of other renewable power resources, such as salinity gradient, marine hydrokinetic, and ocean thermal energy conversion, appear promising; however, most of these technologies are yet to commercialize and establish their potential.

Waste heat from power plants is an abundant source of low-cost energy for thermal desalination technologies, which can operate effectively at low temperatures. It produces distilled-quality product water with around 1 ppm TDS from seawater without any chemical pretreatment of the feed and uses less expensive construction materials such as epoxy-lined carbon steel for the evaporator shell. The overall efficiency of the low-temperature evaporation (LTE) process is low due to low-temperature operation. Therefore, the capacity of these systems tends to be on lower side. Nevertheless, LTE shows good potential for small-capacity applications such as supplying the in-house water requirements of coastal industries and commercial centers.

Waste heat is thermal energy rejected at low temperature by thermal and industrial plants. In some cases, waste heat can be directly used as heat input for the desalination system. In other cases, its use for desalination may require modification of the upstream process to account for differences in temperature or heat load. While the potential for use of such low-temperature heat is substantial, its systematic utilization as energy input for desalination is case sensitive.

Improvements to either energy-production or water-production technologies can help to reduce the overall contribution of desalination processes to global warming. Desalination systems themselves have seen steady improvements in energy efficiency in recent decades. Research and development have contributed to a progressive reduction of the energy demand. In terms of its cost and carbon footprint, seawater reverse osmosis has improved considerably over the past few decades.

As mentioned, the conventional seawater reverse osmosis process consists of a semipermeable membrane capable of selectively separating pure water from seawater. The RO membrane is based on a non-porous thin-film composite (TFC) design in which a dense polymer skin provides the active-site architecture responsible for ion rejection. The development of advanced membranes with high water permeability and salt rejection has the potential to reduce energy consumption and save capital and operational costs, contributing to a reduction in carbon footprint. Nanostructured high-flux membranes have the potential to significantly enhance performance. The associated high permeability can result in a decrease in the pressure needed to drive permeation, thereby reducing energy consumption, potentially to as low as 1.5 kWh/m^3, and increasing permeability and productivity.

Mixed-matrix membranes (MMMs), in which a filler material is embedded within a polymeric matrix, are being explored. Separation performance can be tailored and new functionality added to membranes for desalination applications.[10] MMMs use nanomaterials as a filler material and are known as thin-film nanocomposite (TFN) membranes. These membranes incorporate metal nanoparticles (such as gold, silver and iron oxide), zeolite, carbon nanotubes, TiO_2, graphene and polymerizable surfactants, etc. In addition, inorganic TFN membranes are being explored that use nanomaterials, but not mixed with polymers, to increase water permeability. For example, graphene has the potential to be developed as a high-permeability desalination membrane with better salt rejection than a conventional membrane. The new membranes currently in development use carbon nanotubes, graphene and aquaporin, among others. The common objective of all this membrane development work is high-flux operation, several times higher than with conventional membranes. The drawback of the new RO membrane is that the conventional UF pretreatment membrane allows small organic molecules that may quickly plug the high-flux membranes. This, in turn, increases the required feed pressure, adversely affecting the anticipated low carbon footprint of future desalination technology. There is a need to develop a next-generation ultrafiltration membrane, taking into account current limitations.

In remote areas which may lack an electrical grid and proper infrastructure, desalination plants are typically powered by a diesel generator, which is both expensive and polluting. A renewable energy-powered desalination plant can be the optimal solution in such cases. A reverse osmosis desalination plant coupled with photovoltaics is a promising option for small-scale desalination in remote areas, although a major challenge is the plant's ability to accommodate variations in the solar radiation, which is directly linked to variation in the water production. The intermittent power input received by the plant requires the system to adjust its settings. Earlier reverse osmosis desalination systems combined a photovoltaic array and batteries to store energy. Battery-based systems are relatively expensive and have limited life. In recent years, photovoltaic-powered reverse osmosis systems without batteries have been the subject of research. The following challenges still need to be overcome:

(i)   Design challenges
 • intermittent operation and therefore need for customized design
 • large space requirement for the energy supply system (solar PV modules)
 • availability of energy recovery device for small unit
 • limited operational experience
(ii) Logistical challenges
 • local availability of chemicals (pretreatment, anti-scalant, post-treatment)
 • water quality monitoring (off-site laboratory)
(iii) Operation and maintenance (O&M) challenges
 • limited availability of experienced and trained personnel

The challenge is to develop a suitable, sustainable desalination system to provide drinking water, with a strong focus on reliability, high availability and minimal maintenance, even in extreme conditions.

In terms of technology readiness level (TRL), process improvements for energy efficiency, hybrid desalination technologies and advanced pretreatment technologies can play an important role in the near future, producing a high impact and high TRL. Salinity gradient energy recovery, forward osmosis and membrane distillation are associated with a relatively low TRL. Four areas are associated with high TRL, high-impact low-carbon desalination: (i) PV-RO, (ii) wind-RO, (iii) CSP-thermal desalination hybrids, and (iv) optimized power–water cogeneration. Salinity gradient power is relatively low priority in terms of impact and TRL. Nuclear-RO combinations (either grid driven or perhaps standalone with small modular nuclear reactors below 300 MWe) are also recognized as having high potential impact on GHG emissions, and a generally high TRL. Nuclear-thermal hybrids have a similar impact. Further research is recommended to explore the long-term reliability of desalination systems when operated intermittently with renewable energy.

5.6   RECOVERY OF VALUABLE MATERIALS FROM BRINE EFFLUENT

Effluent rejected from sea water desalination plants or water treatment plants in industry may contain a number of valuable materials and is a source of many chemicals. Some of the elements are both very scarce and expensive. There is thus a strong motivation to recover them from the reject brine of a desalination plant or from industrial effluent. This adds value to a desalination plant or industry, as well as making it more eco-friendly.

Recovery of uranium (U) and other valuable materials from the concentrate brine of desalination plants appears promising. The recovery of these materials is desirable not only in order to meet demand but also in reducing the cost of desalinated water.

R&D work on recovery of uranium from the effluent of desalination plant has been pursued through the ionizing radiation and chemical synthesis routes.[11-13] Electron beam irradiation has been used to provide radiation grafting of acrylonitrile (ACN) on a non-woven polypropylene (PP) fiber substrate. Recovery of uranium at milligram level i.e., 1,000 µg /g of poly (amidooxime) (PAO) in 52 days at 25°C in laboratory conditions has been reported by Japanese researchers. It was observed that vanadium is also collected on the adsorbent. The potential of polyhydroxamic acid (PHOA) sorbent for uptake of uranium from seawater was also evaluated chemically and a concentration factor of over 190 was obtained when the resin, filled in a porous bag, was dipped in seawater for a period ranging from 10 to 30 days. About 4.5 billion tons of uranium is present in sea water, a thousand times more than what is known to exist in uranium mines. Since its concentration is extremely low, harvesting uranium from the sea is a challenging task.

5.7   BRACKISH WATER DESALINATION

Water containing salinity in the range of 1,000–20,000 ppm total dissolved solids (TDS) is generally known as brackish water. Different types of desalination processes are used to bring down the salinity of water. Brackish water desalination using a membrane process to get drinking quality water from saline water is practiced widely. There are several advantages of brackish water desalination by membrane process, such as low specific energy requirement and low-pressure membranes. It has short start-up and shutdown times. It can be easily integrated with renewable energy sources including solar, wind, tidal, etc. Technological innovations have brought down the specific energy consumption in brackish water RO considerably to about 1 kwh/m³. The incorporation of advanced membrane-based pretreatment technology such as microfiltration (MF) and ultrafiltration (UF) has the potential to increase plant recovery, reduce consumables required for operation and maintenance, reduce membrane cleaning frequency and increase membrane life.

The selection of appropriate technology and plant capacity is an important and challenging task, particularly in remote rural areas. At times, it is advisable that capacity should be restricted to small or medium size, depending on yield and the availability of brackish water in the area compared to the requirements. Small- and medium-capacity truck- or barge-mounted desalination plants may be useful for supplying purified water on a need basis to coastal and remote rural areas. Low-cost rugged and affordable devices need be developed for installation at domestic and community levels. Development of small solar desalination units for remote locations appears promising. Disposal of concentrate and reject water is a challenging issue for inland desalination plants. Inappropriate discharging may pollute aquifers and groundwater as well as the environment over a period of time.

Electrodialysis is quite promising for low-salinity brackish water desalination. It is an energy-intensive technology for sea water desalination. Membrane distillation, forward osmosis and humidification-dehumidification are in different stages of development and yet to be deployed for large-scale applications. Hybrid systems combining two or more desalination technologies have the potential to increase water recovery and consequently reduce brine management costs or lower specific energy requirements.

There is a definite need for collaboration between industry and academia to develop and build efficient, cost-effective desalination systems. Some of the research areas to be pursued are as follows:

1. Development of new-generation membranes (such as nanocomposite membranes) having high flux, long life, biodegradable characteristics, temperature-tolerant properties and resistance to biological fouling.
2. Spent membrane management.
3. Advanced membrane-based pretreatment methods.
4. Low-grade/waste heat utilization as energy input for seawater desalination.
5. Advanced heat transfer surfaces for thermal desalination systems.
6. Hybrid desalination systems.
7. Recovery of strategic and precious materials from brine rejected from the desalination plant.
8. Development of innovative seawater intake and outfall systems.
9. Innovative desalination processes.
10. Reduction in energy consumption.
11. Development of eco-friendly integrated zero-liquid-discharge systems.

Some of the constraints normally encountered in remote areas pose certain challenges. Power supply and voltage fluctuation issues are significant concerns in many remote regions. Availability of power supply varies at times, and even the available power supply is not regular, with crippling voltage fluctuations and sudden power cuts. The inaccessibility of remote areas poses problems in case of equipment failure where skilled human resources and spare parts may not be readily available. Designs should therefore be robust and simple, and ensure that minimum human interface is called for. Sustainability in remote rural areas is challenging. Awareness as well as local (including financial) participation must be ensured. For sustainability, it is essential that the demand should be generated from the user side. Innovative and robust business models are needed. Technology selection, affordability, acceptability and accessibility are the important parameters for strategy development.

5.8 ALTERNATE TECHNOLOGIES

5.8.1 SOLAR AND WIND-POWERED DESALINATION

Solar thermal energy may be used in thermal desalination plants such as MSF and MED, while solar photovoltaics (SPVs) are suitable for membrane-based plants. The use of direct solar energy for desalination has been investigated for quite some time. Considerable work was carried out during World War II on small solar stills for use in life rafts. Solar energy is free and abundantly available. Although the technology has huge potential, work is still needed to reduce the high installation costs and large space requirements. Several desalination units have been set up using energy derived from solar, such as wind energy and ocean thermal energy. While initial investment costs are high for SPV plants, improvements in efficiency and mass-production of SPV cells can substantially reduce them. The main focal points in photovoltaic cell research are efficiency increase, reduced manufacturing costs and the search for other materials as GaAs, CdS, CdTe and $CuInSe_2$ (CIS). In cells of GaAs and its alloys, such as $GaInP_2$, efficiencies higher than 30% (ultra-high efficiency) have been achieved. There is a trend towards decreasing the cost of solar electricity. SPV-powered reverse osmosis units appear an attractive option due to depletion of fossil-fuel reserves. Other concepts such as solar ponds providing energy to MED/ MSF are also gaining attention.

Wind energy-powered small-scale desalination plants have good potential in locations where water transport costs outweigh renewable energy source expenses. It must be kept in mind that the power output fluctuates with variations in wind speed. The equipment has a typical wind-to-mechanical maximum efficiency of about 45%, with the average about half this value. There is scope for development in wind turbines with advances in control strategies and improvements in energy storage systems.

Wind-SPV hybrid-powered desalination systems are an attractive alternative since the desalination plant can operate even when one of the sources is not of the required intensity.

Coupling of desalination units with non-conventional renewable energy sources is an attractive proposition in remote areas. Using solar distillation on individual roof-tops, the requisite quantity of drinking water for a small household can be produced from a few liters of brackish water. The concept is similar to solar water heaters or solar power panels. Salt manufacturers can incorporate solar distillation instead of open evaporation. This would help in brine management.

5.8.2 MEMBRANE DISTILLATION

Membrane distillation (MD) is a separation process, in which only vapor molecules are able to pass through a porous hydrophobic membrane.[14] It is a thermally driven process. The membrane distillation system is driven by the difference in vapor pressure between the porous hydrophobic membrane surfaces. It has good potential as a thermally driven membrane process for desalination. Energy consumption in MD and the effect of its operating parameters require further investigation, as limited research data is available in this area.

The materials used in membrane distillation are quite different from those for pressure-driven membrane processes. In the case of membrane distillation, the membrane should have low resistance to mass transfer and low thermal conductivity to prevent heat loss across the membrane. It should also have good thermal stability and high resistance to chemicals. The water vapor from the hot feed (brackish/sea water) permeates across the membrane due to the thermal gradient and condenses on the other side, enabling the collection of relatively pure water. Hence, the commonly used materials in membrane distillation are polytetrafluoroethylene (PTFE), polypropylene (PP) or polyvinylidene fluoride (PVDF) polymer. The membrane distillation process has various applications, such as desalination, wastewater treatment, etc.

The thickness of the membrane is significant in membrane distillation systems. Permeate flux is inversely proportional to membrane thickness. The permeate flux is reduced as the membrane thickness is increased. Resistance to mass transfer increases, while heat loss is also reduced as the membrane thickness increases. Membranes with pore size in the range of 100 nm to 1 μm are normally used for MD. A large pore size gives high permeate flux. On the other hand, small pores avoid liquid penetration. This implies a requirement for optimum pore size range on a case-to-case basis.

Membrane distillation has several advantages, such as lower operating temperatures than conventional thermal processes. The hydrostatic pressure is lower than that used in pressure-driven membrane processes such as reverse osmosis. Membrane characteristics are less exacting, hence membrane material need not be expensive. Membrane distillation gives a high separation factor. The membrane pore size is relatively bigger than for other membrane separation processes. It is relatively more resistant to fouling. Membrane distillation systems can be combined with other separation processes to give an integrated hybrid membrane separation system, such as with a UF or RO unit.[15] The system has potential to utilize alternative energy sources, such as solar energy.[16] It appears quite promising for desalination of brackish water and seawater and has good potential for the removal of organic and heavy metals from aqueous solution[17] and wastewater. It can be used to treat radioactive waste, where the product can be safely discharged to the environment.[18] However, membrane distillation faces some major challenges such as low permeate flux. Also, trapped air within the membrane may introduce an additional mass transfer resistance, which also limits the permeate flux. The heat lost by conduction is quite significant. Mass transfer in membrane distillation is controlled by three basic mechanisms: (i) knudsen diffusion, (ii) poiseuille flow (viscous flow) and (iii) molecular diffusion. This increases mass transfer resistance resulting from transfer of momentum to the supported membrane (viscous), collision of molecules with other molecules (molecular resistance) or with the membrane itself (Knudsen resistance). The mass transfer boundary layer resistance is generally negligible.[19] Similarly, the surface resistance is insignificant, because the surface area of the membrane distillation is small compared to the pore area. On the other hand, the thermal boundary layer is considered to be the factor limiting mass transfer.[20]

Different types of configurations are used to separate aqueous feed solution using a microporous hydrophobic membrane.

5.8.2.1   Direct Contact Membrane Distillation (DCMD)

In direct contact membrane distillation, the feed is at higher temperature as it is in direct contact with the hot membrane surface. Evaporation takes place on the feed-membrane surface. The vapor is driven to the permeate side due to the pressure difference across the membrane. It condenses in the membrane module. The feed in liquid form cannot penetrate the membrane because it is hydrophobic. DCMD is a simple configuration widely used in desalination processes. The main drawback of this configuration is the heat loss by conduction.

5.8.2.2   Air Gap Membrane Distillation

The feed solution in the air gap membrane distillation (AGMD) process is in direct contact with the hot side of the membrane surface. In AGMD, stagnant air exists between the membrane and the condensation surface. The vapor needs to cross the air gap to condense over the cold surface inside the membrane cell. The advantage of this design is reduced heat loss by conduction. However, the disadvantage is additional resistance to mass transfer due to the air gap. This type of configuration is suitable for desalination and for removing volatile compounds from aqueous solutions.

5.8.2.3   Sweeping Gas Membrane Distillation (SGMD)

In sweeping gas membrane distillation (SGMD), inert gas is used to sweep the vapor on the permeate side of the membrane to condense outside the membrane module. In SGMD, as in AGMD, there is a gas barrier to reduce heat loss. However, the gas barrier in SGMD is not stationary, so it enhances the mass transfer coefficient. This configuration is useful for removing volatile compounds from aqueous solution. Its main disadvantage is the requirement for a large condenser because a small volume of permeate diffuses in a large sweep-gas volume. AGMD and SGMD can be integrated in a process called thermostatic sweeping gas membrane distillation (TSGMD). The inert gas, in the case of TSGMD, is passed through the gap between the membrane and the condensation surface. Part of the vapor is condensed over the condensation surface (AGMD) and the remainder is condensed outside the membrane cell by the external condenser (SGMD).

5.8.2.4 Vacuum Membrane Distillation (VMD)

In VMD, a pump is used to create and maintain the desired vacuum in the permeation side of the membrane. Condensation occurs outside the membrane module. The heat lost by conduction is negligible, which is considered a significant advantage. This type of MD is used to separate aqueous volatile solutions.

5.8.3 Capacitive Deionization

Capacitive deionization (CDI) is a promising water treatment technology which uses electrophoretic driving force to achieve desalination. Direct current is used to remove ionized species from the feed stream, while continuously regenerating the resin pack. Capacitive electro-deionization devices are normally used for removing the last traces of dissolved ions to produce ultrapure water using resins and a polymeric ion-exchange membrane. Extension of this concept to the desalination of brackish water could be useful, as it may give high water recovery rates. In CDI, ions are adsorbed on the surface of porous electrodes by applying a low voltage (1.0–1.6 V DC) electric field. The compact and modular system removes dissolved salts without the need for chemical regeneration, eliminating waste disposal problems. Positively charged ions such as calcium, magnesium and sodium are attracted by negative electrodes. Negatively charged ions such as chloride, nitrate and sulfate are attracted by positive electrodes. In CDI, no additional chemicals are required for regeneration of the electro-sorbent. Eliminating the electric field allows ions to get desorbed from the surface of the electrodes and regenerates the electrodes. There are several types of electrode materials and configurations to enhance performance. Optimized carbon aerogel is a promising electrode material due to its high electrical conductivity, high specific surface area and controllable pore size distribution. The energy requirement of CDI depends on the amount of salt removed. CDI technology is cost competitive with conventional processes at a low feed concentration range (<3,000 mg/L) due to the high cost of CDI modules with increased feed water concentration.

5.8.4 Desalination Through Carbon Aerogel

Carbon aerogel (CAG) an extension of capacitive deionization except that it uses inorganic matrix and nano-material. CAG is a form of carbon with reticulated structure whose pore sizes are in the order of a few tens of nanometers. It acts like a microscopic sponge with high electrical conductivity. It possesses a large surface area inside the pores (about 100–700 m^2/cm^3). If a pair of opposite-charged CAG electrodes are dipped into saline water (brackish or seawater) and a voltage of about 1.0–1.5 V is imposed, the opposite-charged ions are attracted and sorbed on the appropriate electrodes, leaving behind pure water. It is reported to have high capacity for the removal of large quantities of salt. About 2 kWh/m^3 was reported to be the power consumption for desalinating water with from a few hundred ppm to less than 0.1 ppm of dissolved salts. For seawater, the estimated energy consumption is about 5.5 kWh/m^3.

5.8.5 FREEZING

Desalination by freezing is achieved by cooling the saline water to form ice crystals under controlled conditions and melting the ice to obtain pure water. In 1786, Antonio Maria Lorgna was the first person to describe a working method for desalination by freezing.[21] Extensive work was carried out during the 1950s and 1960s. The process has several advantages such as low energy consumption as well as low scaling and corrosion probabilities due to low-temperature operation. The freezing-based separation process involves handling ice and water mixtures. If the surface of the ice remains coated with brine, it results in high salinity of the product water. To avoid this, the ice is washed with pure water. The amount of pure water required for washing is important and should be as low as possible. Pilot plants for freezing process-based desalination systems have been established; however, they have still to achieve commercial success.

5.8.6 WATER HARVESTING FROM AIR

In this type of system, water is extracted from humid air by condensing water using vapor compression refrigeration. Brackish water also can be used in a closed-loop system for recovery of fresh water. More research and development efforts are required to enhance efficiency of these processes.

5.8.7 SUBMARINE DESALINATION

This is also known as reverse osmosis deep sea system (RODSS). The idea is to carry out desalination on the seabed (e.g., 600 m deep), with a hydrostatic head replacing the conventional RO seawater-resistant high-pressure pump. Only freshwater has to be pumped up to the surface reservoir. Special arrangements are required to protect the modules and cables against high external pressure. Deep-bed seawater is free from impurities, which reduces biofouling.

5.8.8 THERMOCLINE-DRIVEN DESALINATION

Thermocline-driven desalination is a low-temperature distillation process for seawater desalination that makes use of the ocean temperature difference (thermocline) existing between the surface and the deep seawater (hundreds of meters deep). This desalination system utilizes a flashing mechanism. The technology is eco-friendly; however, the equipment is quite large due to the low-temperature operation and low heat transfer coefficient, which limits the capacity of the desalination plant.

5.8.9 UTILIZATION OF MUNICIPAL SOLID WASTE

Controlled incineration of municipal solid waste (MSW) with flue gas emission can generate steam that can drive a turbine. It is estimated that 188 L of freshwater can be produced by burning one kg of MSW.

5.9   COST REDUCTION THROUGH TECHNOLOGICAL INNOVATIONS

In the case of water technologies, cost is quite sensitive to:

- User-driven factors like capacity, quality etc.
- Technological components
- Environmental aspects
- Logistics

The contribution of each of the above aspects varies with site-specific parameters. The techno-economics of large- and small-capacity plants depend on local factors including logistical/infrastructural support and environmental factors. Cost-reduction strategies are essentially technology driven.

The investment and total water costs are primary parameters used by decision makers to select the appropriate desalination technology for a project based on plant size, energy requirements, contract type and other factors, such as environmental considerations. However, cost depends on several parameters, and the accuracy of a cost estimate depends on transparent and high-quality software packages for accurate results. Cost data reports are generally not very consistent for different technologies or similar-sized facilities, because they are site specific and depend on local factors. The complexity and content of cost estimation models needs to be decided. Water cost estimation methodologies need to identify and specify all the parameters that contribute to desalination costs in order to develop a structured and transparent procedure for estimating desalinated water cost for any facility. These estimates are needed for project planning and budgeting as well as for feasibility studies. Desalination costs are falling for all technologies due to technology innovations, particularly in the last decade, with significant cost reductions occurring in RO technology. The reduction in RO treatment cost is due to the growth rate, plant capacity, competition with other technologies and improvements in RO systems. Better process designs, better membranes and materials, and lower energy consumption also contribute to cost reduction. Simplicity and flexibility of project bids are also important factors contributing to cost reduction. However, desalination costs may not continue to decrease at the same rate in the near future, despite continued improvements in the existing technologies. Equipment, raw materials and energy costs are rapidly rising, which may impact future capital and operating costs. Environmental guidelines are also becoming more stringent. Sustainability aspects are becoming more and more important.

5.10   CASE STUDIES

5.10.1   SEAWATER DESALINATION

Seawater desalination provides an additional source of fresh water, enhancing fresh water availability in coastal areas. Co-location of a desalination plant and power plant has the benefit of sharing common resources and infrastructure facilities. As many of the water-scarce areas lie in coastal regions, there is great potential for seawater desalination.

5.10.1.1 Nuclear Desalination Using Hybrid Technology

Nuclear energy has potential to deal with the carbon footprint, as it provides low-carbon desalination. It can take the form of heat/electricity producing fresh water from seawater. The pressurized heavy water reactor (PHWR)-based nuclear power plant uses natural uranium as fuel and heavy water as moderator. It operates at about 4 megapascal (MPa) steam pressure and 250°C steam temperature. As the enthalpies of steam available at the entrance to the high-pressure (HP) turbine of a pressurized heavy water reactor-based nuclear power plant (NPP) are lower than with a fossil fuel-based conventional power plant, the specific steam consumption in the nuclear power station is higher. This means more steam is available from nuclear power plants that could be gainfully utilized for thermal desalination (Table 5.4).

A nuclear desalination plant based on hybrid technology was developed by the Bhabha Atomic Research Centre (BARC) in India. It is known as the Nuclear Desalination Demonstration Plant (NDDP). NDDP consists of a hybrid MSF-RO desalination system of 6.3 million liters per day (MLD) capacity (i.e., 4.5 MLD MSF and 1.8 MLD RO) coupled to 2 × 170 MWe PHWRs at Madras Atomic Power Station (MAPS) at Kalpakkamin, India (Figure 5.6). The seawater, steam and electrical power requirements for the desalination plants are met from MAPS and are around 1.5%, 1.0% and 0.5% of MAPS' availability (Figure 5.7).[22]

TABLE 5.4

Salient Features of Nuclear and Fossil Power Sources for Desalination

S. No.	Characteristics	Nuclear Power Station	Thermal Power Station
1	Type	PHWR	Coal based
2	Steam pressure (MPa)	4	13
3	Steam temperature (°C)	250	535
4	Steam enthalpy at high-pressure turbine inlet (kJ/kg)	2,800	3,470
5	Specific steam consumption (kg/kWh)	6	3.4
6	Steam pressure at low-pressure turbine outlet (MPa)	0.01	0.01
7	Steam temperature at low-pressure turbine outlet (°C)	45	45

MAPS   MSF   UF-RO

FIGURE 5.6 Nuclear desalination plant at Kalpakkam, India.

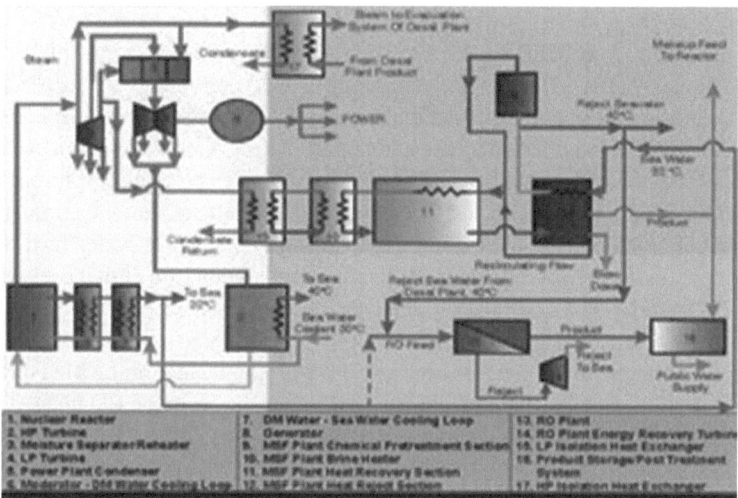

FIGURE 5.7 MSF-RO hybrid nuclear desalination plant coupled to 170 MW(e) PHWR Station at Kalpakkam, India.

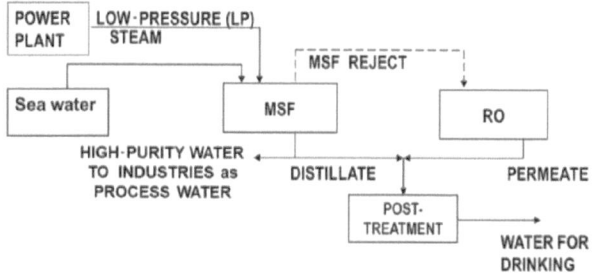

FIGURE 5.8 Advantages of hybrid system.

The hybrid technology has provision for redundancy, utilization of streams from one to another and production of two qualities of water for best utilization (Figure 5.8). RO operates with the help of electricity, whereas MSF uses low-pressure steam. Two types of desalinated water are produced. Water produced from the MSF desalination system is of distilled quality which is good for high-end industrial use. Water produced from the RO desalination system is of potable quality. The two can be blended if required.

The requirements of seawater, steam and electrical power for the desalination system are shared with MAPS. The RO plant incorporates necessary pretreatment and an energy recovery system. It operates at relatively lower pressure and employs fewer pretreatment chemicals because of the relatively clean feed seawater from the seawater outfall of nuclear power stations. The potable water produced is supplied to nearby areas. The MSF plant is designed for higher top-brine temperatures with a GOR of 9:1

and utilizes less pumping power due to the long-tube design. A part of the distilled water produced from MSF is supplied to MAPS for high-end applications such as demineralized water. The hybrid technology is very promising in urban areas of coastal regions. Co-location of desalination and power plants has the benefit of sharing resources such as common intake and outfall of sea water and other infrastructural facilities. The concept of hybrid desalination economizes product cost through a mix of different desalination processes. The concept takes into account the product quality requirements for different uses, such as providing high-quality boiler feed make-up water or process water for industrial applications and potable water to the people in the neighboring areas. The desalination plant can meet the water needs of around 45,000 people at 140 liters per capita per day (LPCD). There is provision for augmentation of product water capacity by blending the low TDS product water of the MSF plant with brackish groundwater/ moderate-salinity permeate water from the RO plant. Thus, the hybrid MSF-RO desalination technology has the following advantages:

(i) The RO plant continues to be operated to provide the minimum quantity of desalinated water for drinking purposes during shutdown of the power station.

(ii) The return cooling seawater from the MSF plant can be utilized (after blending with raw sea water) as feed for the RO plant for its enhanced throughput.

(iii) Some of the high-purity product water from the MSF plant is used as the make-up process water after necessary polishing for the power station as well as in the nearby areas for high-end requirements.

(iv) Mixing of the product water from both RO and MSF systems provides drinking water of the requisite quality.

This technological innovation has the potential of utilizing nuclear energy for sea water desalination in water-scarce coastal areas and large coastal arid zones. Dual-purpose plants (power-cum-desalination plants) have inherent design strategies to optimize thermodynamic efficiency alongside economic optimization.

The steam at around 3.5 bar pressure is tapped for the MSF plant at the NDDP from the spare opening in the cold reheat line after HP turbine exhaust from both the nuclear reactors. Moisture from the steam is removed through a moisture separator and steam is sent to an intermediate isolation heat exchanger (IHX) to produce process steam (using demineralized water) for the brine heater of the MSF plant. The condensate from the isolation heat exchanger is sent back to the deaerator section of the power station, while the condensate from the brine heater is returned back to the isolation heat exchanger. Adequate monitoring and control provision is made for isolation of the steam supply in case of reactor or desalination plant shutdown. A separate steam source coming directly from the reactor is utilized for ejectors of the MSF plant after passing through another isolation heat exchanger. Technical specifications of MSF plants are given in Table 5.5.

The pretreatment scheme for the MSF plant involves acidification, vacuum deaeration for control of O_2/CO_2 concentration, pH control by alkali neutralization and antifoam dosing. The pretreatment for the RO plant involves ultrafiltration membrane-based systems. Operating parameters of the RO plant are given in

TABLE 5.5
Technical Specifications of 4.5 MLD MSF Plant

1.	Product water output (m³/h)	187.5
2.	Product water salt content (ppm)	2–5
3.	Sea water requirement (m³/h)	
	i) Total (cooling)	1,500
	ii) Make-up feed	375
4.	Max. recirculation brine temperature (°C)	121
5.	Blow-down brine temperature (°C)	40
6.	Concentration ratio of blow-down brine	2
7.	Steam consumption	
	i) Heating in the brine heater (tonnes/h)	20.6 at 2.8 bar
	ii) Steam jet ejectors (tonnes/h)	0.4 at 7 bar
8.	Performance ratio (kg water production per kg steam input to the brine heater)	9
9.	Overall electricity consumption (kWh/m³)	2.5

TABLE 5.6
Design Parameters of 1.8 MLD Sea Water Reverse Osmosis Desalination Plant

1.	Product water	
	a) Output (m³/h)	75
	b) Salt content (ppm)	200–500
2.	Sea water	
	a) Requirement (m³/h)	215
	b) Total dissolved solids (ppm)	35,000
3.	% Recovery	35
4.	Average salt rejection (%)	99
5.	Operating pressure (bar)	45
6.	Operating temperature (°C)	32–36
7.	Overall energy consumption (kWh/m³)	5

Table 5.6. The RO plant operates at relatively lower pressure to save energy, employs fewer pretreatment chemicals (because of relatively clean seawater from MAPS outflow) and aims for longer membrane life. The potable water produced is supplied to nearby areas.

The cost of desalinated water is site specific and depends on a variety of parameters. The capital cost consists of the infrastructure requirements at a given site and the cost of desalination plant, detailed engineering, consultancy and finance. The operating cost consists of energy, consumables, manpower and miscellaneous overheads besides amortization cost. Capital cost is very sensitive to capacity and local industrial infrastructure, particularly for thermal plants. The scale-up advantage of thermal plants is very significant. In addition to this basic advantage of nuclear power plant over a conventional thermal source, nuclear energy has a few other added advantages: (i) it is cheaper for places situated a long way from coal-pit heads; (ii) it is eco-friendly because it is free from greenhouse gas emissions; and (iii) it assists in sustainable growth of a total energy generation program.

The PHWR system provides an added opportunity in the form of a very significant additional source of thermal energy that can be effectively utilized for desalination by low-temperature evaporation (LTE)-type desalination systems in water-scarce coastal regions. About 40 MW(th) and 100 MW(th) of waste heat is available in the heavy water moderator system of the 220 MW(e) and 500 MW(e) capacity PHWR respectively. A significant part of this waste heat can be utilized for seawater desalination, producing distilled water for in-house consumption in the nuclear power station. It is possible to produce up to 1.0 MLD distilled-quality water by sea water desalination using the waste heat of the heavy water moderator from a 500 MW(e) coastal PHWR to meet its make-up DM water requirement. LTE technology for producing distilled water from seawater by using the waste heat has been developed in India.

A system has been developed to conduct practical demonstrations of this nuclear desalination technology based on LTE. A 30,000 LPD LTE desalination system was integrated with a PHWR-based nuclear research reactor at BARC Trombay, India. Waste heat was used to produce demineralized (DM)-quality make-up water (Figure 5.9). The desalination plant was designed, installed and integrated with a nuclear reactor system. It demonstrated the utilization of the waste heat of a nuclear research reactor for sea water desalination. The desalinated product water was used as the high-quality make-up water for the reactor. This concept of using the waste heat has resulted in water demineralization cost saving, due to the high purity of desalinated water as well as the conservation of water because seawater is a new source of water.

The operating parameters of the LTE desalination plant are given in Table 5.7.

Among the added attractions of the LTE process are:

i)   It is eco-friendly and does not require chemical treatment of feed water.
ii) It replaces the cation and anion exchangers and eliminates the requirements for regeneration chemicals.
iii) As it uses sea water, the raw water which is used as feed for the DM plant could be released to local people for use.

FIGURE 5.9 LTE Seawater Desalination Plant utilizing Waste Heat of Nuclear Research Reactor.

TABLE 5.7
Operating Parameters of a 30,000 L/day LTE Desalination Plant Using Waste Heat

Product water output (m³/h)	1.25
Sea water salinity (ppm)	35,000
Product water quality (ppm)	2–5
Total sea water cooling requirement (m³/h)	144
Feed sea water requirement (a part of total requirement) (m³/h)	4.5
Hot water temperature (°C)	53–65
Sea water boiling temperature	41
Operating pressure (abs bar)	0.08
Waste heat requirement (MWth)	1
Electric power consumption (kW(e))	
a) if sea water is available at 6 bar	2.25
b) if sea water is available at 1 bar	15

5.10.1.2   Multi-Effect Distillation Desalination in a Petrochemical Complex

Reliance Industries Limited (RIL) on India's west coast is a fully integrated manufacturing complex housing a petrochemical refinery, captive power plant, port and terminal facility. It is one of the largest grassroots refineries intended to provide backward integration to petrochemical plants by producing feed stock and other basic petrochemical products, besides producing liquefied petroleum gas (LPG) and diesel to meet the ever-growing demand for these products. It was built to process crude oil and to cater to the demand for petrochemical products in north-west India. The site is adjacent to the town of Jamnagar in Gujarat State alongside the Gulf of Kutch. Total area of the complex is 7,500 acres. This coastal land is prone to drought every alternate year, with average total rainfall of only 61 mm/year. The water source is becoming a major cause for concern. To cater to the industrial requirements of the process as well as drinking water for about 10,000 working people was a major concern. Depending on rainfall as the only source of water was not a feasible option. The nearest area for sourcing and transporting water was 350–500 km from the plant location. To bring water through pipelines from this place was not advisable. The water requirement of the refinery complex was of the order of 36 MLD (1,500 m³/h). For a refinery located on the coast in a drought area to cater for such a large requirement, it was decided by the management to convert sea water into the desired quality of water. As low-pressure steam is available in the refinery complex, it was further decided to install a thermal desalination plant for producing fresh water from seawater.

A multi-effect desalination plant with an installed capacity of 48 MLD (2,000 m³/h) producing high-purity distilled-quality water (5 ppm TDS) was thus installed at the Jamnagar complex. The desalination plant is based on the MED process and was supplied by Israel Desalination Engineering (IDE), Israel. It has a series of horizontal tubes and falling film-type evaporators. The train consisting of a total of eleven evaporator units was divided into three groups: the hot group (effect nos 1, 2 and 3), the intermediate group (effect nos 4, 5, 6 and 7) and the cold group (effect nos 8, 9, 10 and 11). A final condenser was connected at the end of the train. The cold group is towards the condenser side and the hot group is at the opposite end. Sea water first enters the condenser and is pumped to the cold group, the intermediate group and lastly to the hot group through a series of pumps. Finally, it is rejected from the hot group of effects as brine which is then sent back to the sea. The flow of steam is from first effect to eleventh effect. When sea water is heated by low-pressure steam and evaporated due to transfer of latent heat of steam, an equivalent amount of vapor is generated. The same vapor is used as a heating medium to evaporate sea water in the next vessel or effect. The operation is carried out in multiple effects with simultaneous evaporation and condensation. Performance of the MED desalination plant is monitored through GOR, which is defined as the ratio of the amount of distillate produced to the amount of steam used.

Pretreatment of feed water is an important step in the distillation of seawater and is a must to overcome problems related to scale formation, corrosion and organic growth. The pretreatment facility should have provision for filtration, chlorination, scale control measures and deaeration. The first two steps are provided externally at the sea water intake facility, whereas the latter two steps form part of the desalination plant. Table 5.8 gives the basic operating parameters of the MED desalination plant installed at Reliance, Jamnagar.

As a result of successful operation and performance, the desalination capacity was added in 2005, 2008 and beyond, augmenting fresh water production. It has exceeded the 160 MLD capacity based on MED technology.

TABLE 5.8
Basic Operating Parameters of MED Desalination Plant Installed in a Petrochemical Complex in India

Make	IDE Technologies Ltd., Israel
Model	MED-12000
No. of Units	Four
Type of Unit	Horizontal tube, falling film, multi-effect desalination unit (with vapor cycle)
No. of effects	11 heat recovery evaporators 1 heat rejection condenser
Distillate capacity (cubic meters per day (each unit))	12,000
Distillate quality (ppm TDS)	5
Feed to evaporator (cubic meter per day (each unit))	1145
Feed water salinity (ppm TDS)	40,000
Sea water inlet temperature (°C)	35
Steam pressure (bar)	3.5
Product out temperature (°C)	40
Brine out temperature (°C)	40
Brine concentration (ppm TDS)	70,000
1st effect brine temperature (°C)	70

5.10.2 BRACKISH WATER DESALINATION IN RURAL AND REMOTE AREAS

Kisari village (population about 1,500) is 30 km away from Jhunjhunu district in Rajasthan state of India. The village is situated in a remote area where no other source of drinking water is available, except the saline water in the well which is used by villagers for drinking, cooking and domestic purposes. The village is identified by the local government as one where an appropriate desalination unit could be installed to provide clean drinking water for the villagers.

An RO unit of 1,800 LPH capacity has been designed and installed. The RO unit has three pressure vessels. Each vessel contains two membrane elements. Each membrane element is 100 mm diameter and 1,000 mm long. The feed saline water from the well is drawn with a pump, and enters a sand filter for removing suspended solids and turbidity in raw water. Thereafter, the raw water enters into micron filters of 10 μm and 5 μm size. Subsequently, the feed water with a turbidity of less than 1 NTU is pumped into the first and second vessels of the RO module at 10–15 bar pressure. The concentrated water emerging from the first and second vessels becomes the feed for the third vessel. Finally, the concentrate from the third vessel is taken out and collected in a tank. The people use the concentrate water, which is about 4,000 ppm saline, for cleaning floors, washing and other non-potable purposes. The permeate water is collected separately from all three vessels and transported to a storage tank by pipeline for further distribution. The villagers use the clean water for drinking and cooking.

The RO unit operates at 55% product recovery at 15 bar operating pressure and 90% salt rejection. Table 5.9 shows the details of feed and permeate water analysis and the rejection by the membrane. The samples show chloride, calcium and magnesium rejection of 90%, 92% and 94% respectively. The presence of fluoride and nitrate observed in the feed water is brought down to permissible levels in the product water.

TABLE 5.9
Kasari RO Plant Performance Results

Parameter	Permeate (mg/L)	Feed (mg/L)	Concentrate (mg/L)	Rejection (%)
Conductivity	342	3,400	7,300	90
Total dissolved solids*	205	2,040	4,380	90
Chloride	50.5	640	1,350	92.1
Sulfate	5.0	90	262	94.5
Potassium	0.637	5.23	9.55	87.8
Magnesium	1.40	25	64.6	94.5
Calcium	7.5	105	215	92.9
Total hardness	9.0	144	290	93.8
NO_3, nitrate	24.2	53.2	79.2	54.5
Nitrate nitrogen (NO_3-N)	5.45	12.1	18.0	55.0
Fluoride	0.134	1.65	3.5	91.9

* Based on conductivity.

When a plant is installed, there may be doubts regarding its sustainability in a rural scenario. However, the Kisari case study has proved that the plant can be operated with the active participation of the local community. When this technology is applied to the provision of drinking water to a rural population, there are associated societal benefits, such as improvement in human health by consuming purified clean water, easy availability, increased awareness of technology in rural sectors, education on self-reliance and the realization that a good and reliable technology can help improve people's quality of life.

REFERENCES

1. Khan, A.H. 1986. *Desalination Process and Multistage Flash Distillation Practice*, Elsevier Science Publishers, UK.
2. Sadhukhan, H.K., Misra, B.M., Tewari, P.K. 1999. Chapter 1: Desalination and wastewater treatment to augment water resources, *Water Management, Purification and Conservation in Arid Climates*, Technomic Publishing, USA.
3. IDA Desalination Yearbook 2016–2017. September 2016, *Global Water Intelligence (Media Analytics)* (UK).
4. Ghaffour, N., Missimer, T.M., and Amy, G.L. 2013. Technical review and evaluation of the economics of water desalination: Current and future challenges for better water supply sustainability. *Desalination* 309: 197–207.
5. Reddy, K.V., Ghaffour, N. 2007. Overview of the cost of desalinated water and costing methodologies. *Desalination* 205: 340–353.
6. Wade, N.M. 1993. Technical and economic evaluation of distillation and reverse osmosis desalination processes. *Desalination* 93: 343–363.
7. *Introduction of nuclear desalination: A guidebook.* 2000. Technical Report Series No. 400, IAEA-TECDOC-400 International Atomic Energy Agency, Vienna (Austria).
8. Low carbon desalination: Status and research, development and demonstration needs. 2016. *Report of a workshop conducted at the MIT Abdul Latif Jameel World Water and Food Security (JWAFS) Lab*, Cambridge, MA, USA.
9. Elimelech, M. and Phillip, W.A. 2011. The future of seawater desalination: Energy, technology and the environment. *Science* 333: 712.
10. Lind, M.L., Ghosh, A.K., Jawor, A. 2010. Tailoring the structure of thin film nanocomposite membranes to achieve seawater RO membrane performance. *Environ. Sci. Technol.* 44: 8230–8235.
11. Prasad, T.L., Tewari, P.K., Sathiyamoorthy, D. 2010. Parametric studies on radiation grafting of polymer sorbents for recovery of heavy metals from seawater. *Ind. Eng. Chem. Res.,* 49: 6559–6565.
12. Pal, S., Prabhakar, S., Thalor, K.L. and Tewari, P.K. 2010. Strategy of deriving 'wealth from waste' from concentrated brine of desalination plant. *Int. J. Nuclear Desalin.* 4: 189–197.
13. Tewari, P.K., Raha, A. and Vinjamur, M. 2012. Integrated zero liquid discharge desalination with seawater flue gas desulphurisation: A new green technology, *Water Today*, 30–42.
14. Al Khudhairi, A., Darwish, N., Hilal N. 2012. Membrane distillation: A comprehensive review. *Desalination* 287: 2–18.
15. Criscuoli, A. Drioli, E. 1999. Energetic and exergetic analysis of an integrated membrane desalination system. *Desalination* 124: 243–249.
16. Blanco Gálvez, J. García-Rodríguez, L. Martín-Mateos. I. 2009. Seawater desalination by an innovative solar-powered membrane distillation system: The MEDESOL project. *Desalination* 246: 567–576.

17. Banat, F.A., Simandl, J. 1998. Desalination by membrane distillation: A parametric Study. *Sep. Sci. Technol.* 33: 201–226.
18. García-Payo, M.C., Izquierdo-Gil, M.A., Fernández-Pineda C. 2000. Air gap membrane distillation of aqueous alcohol solutions. *J. Membr. Sci.* 169: 61–80.
19. Zakrzewska-Trznadel, G., Harasimowicz. M., Chmielewski A.G. 1999. Concentration of radioactive components in liquid low-level radioactive waste by membrane distillation. *J. Membr. Sci.* 163: 257–264.
20. Curcio. E. 2005. Membrane distillation and related operations—A review. *Sep. Purif. Rev.* 34: 35–86.
21. Johnson, W.E., 1993, The story of freeze desalting. *Desalinat. Water Reuse*, 3: 20.
22. Panicker, S.T. and Tewari P.K. 2011. Nuclear energy for water desalination, In *Nuclear Energy Encyclopedia*, Edited by Steven B Krivit, John Wiley & Sons, New Jersey, 65–70.

6 Water Treatment and Purification

6.1 DRINKING WATER PURIFICATION

Standards for desirable and permissible quality of drinking water include the World Health Organisation (WHO) standard, Indian Standard (IS) and European Standard (ES). If the quality of water does not match the drinking water standard due to contaminants present beyond the permissible limit, water requires purification.

6.1.1 COMMON CONTAMINANTS IN WATER

Contaminants adversely affect the quality of water and make it unfit for drinking. Common contaminants in water can be grouped into four major categories: waterborne pathogens, suspended solids, inorganic compounds and organic compounds. These contaminants adversely affect the quality of water and can cause waterborne disease. The basic requirement of drinking water is that its quality should be as per the standard set by World Health Organisation (WHO) or other established standards in the region. Water that has contaminants beyond the permissible limits may need treatment and purification. The drinking water standard may vary from country to country. It is imperative that drinking water should have minerals as well as be clean, bacteria free, odorless and colorless. The total dissolved solids should be within the specified limit and free from toxic substances.

6.1.1.1 Biological Contaminants

Biological contaminants such as pathogens are disease-causing microorganisms. As decreed by WHO, some of the pathogens present in water are a serious risk to human health. Examples include strains of *Escherichia coli*, *salmonella*, *shigella*, *Vibrio cholerae*, *Versinia enterocolitica* and *Campylobacter jejuni*. On the other hand, some of the organisms may cause infection mainly among people with impaired natural defense mechanisms such as the very young, immuno-compromised people and patients in hospitals. For pathogens of fecal origin, water is the main route of transmission. Unhygienic practices during the handling of food, utensils and clothing also play an important role in transmission of pathogens. Humans are typically the main carriers of large populations of bacteria, protozoa and viruses. Pathogens originating from human sources, often from human feces, are called enteric (of intestinal origin) pathogens. The intestines of many domestic and wild animals, their meat, milk and dairy products, are the sources of the bacteria *Yersinia enterocolitica* and *campylobacter*. The persistence of a pathogen in water also influences its transmission to humans. A more persistent pathogen which can survive longer outside the host body is more likely to be transmitted to people. The infective dose of the pathogen determines the number of organisms needed to produce an infection in humans.

Diseases such as diarrhea, dysentery, cholera, typhoid and jaundice are related to microorganisms present in water. There are four major classes of pathogenic organisms related to waterborne diseases. They are bacteria, viruses, protozoa and helminths.

(a)   Bacteria (Prokaryotic)

Bacteria are single-cell prokaryotes (without nucleus) ranging in size from 0.3 to 100 micron (μm) long. Many of these pathogenic bacteria belong to the family *Enterobacteriaceae*. They include the human pathogen; *Salmonella typhi*, which is typically present in all kinds of food grown in polluted environments. Another type of bacterium in this family, *Yersinia enterocolitica* (certain strains) causes acute gastroenteritis with diarrhea. *Y. enterocolitic* bacteria are present in sewage and fecal-contaminated surface water. A special feature of *Y. enterocolitica* is their ability to grow even at low temperatures (say 4°C). Therefore, these organisms can survive for long periods in water habitats. *Shigella*, also part of *Enterobacteriaceae*, causes dysentery in humans and is usually transmitted through direct contact. Other bacteria species of significance but not part of this family include *Vibrio cholerae*, specifically the serogroup 01, which causes cholera, an acute intestinal disease with massive diarrhea, vomiting and dehydration, at times leading to death. Some other pathogenic bacteria include *campylobacter* and opportunistic pathogens such as *Legionella pneumophila* and *aeromonas*.

Escherichia coli, commonly indicating fecal contamination, causes bacterial infections of the intestines where the major symptom is diarrhea. It typically has a length of 3 μm and width of 1 μm. *E. coli* are characterized by their ability to produce potent enterotoxins. Enterotoxins are similar to hormones, and act on the small intestine, causing massive secretion of fluids, which lead to the symptoms of diarrhea. For example, *E. coli* produces a potent enterotoxin that causes both hemorrhagic diarrhea and kidney failure. These diseases may cause death, if left untreated. Waterborne diseases caused by bacterial organisms are cholera, typhoid, paratyphoid, dysentery, diarrhea, tuberculosis, etc.

(b)   Virus

A virus is minute particles containing nucleic acid surrounded by protein and other macromolecules. It lacks many of the cell attributes such as metabolic abilities and reproduction pathways. A virus is smaller than bacteria, ranging in size from 0.02 to 0.3 μm and is known to infect virtually all cells. The pathogenic pathway starts with the attachment of the virion (a virus particle) to a host cell. The virion then penetrates and replicates with the cell, altering the host biosynthetic machinery with its own nucleic acid synthesis.

Most pathogenic waterborne viruses are enteric viruses, which multiply and infect the gastrointestinal tract of humans and animals, before they are excreted in their feces. People infected with any enteric viruses, particularly Hepatitis A, become ill. Infectious hepatitis may cause diarrhea and jaundice, and result in liver damage. Other disease-causing viruses include rotavirus causing gastroenteritis primarily in children, polio virus causing polio and adenovirus causing acute gastroenteritis.

(c)   Protozoan Parasites (Eukaryotic)

Protozoa are unicellular eukaryotic microorganisms that lack cell walls and usually obtain their food by ingesting other organisms or organic particles. Protozoa can infect humans by staying as parasites in the intestines of humans. The most common protozoal diseases are diarrhea and dysentery. *Giardia lamblia* causes an acute form of gastroenteritis. The cyst form is 8–12 μm long and 7–10 μm wide. It is infectious to people by the fecal-oral route of transmission. Their germination in the gastrointestinal tract brings about the symptoms of giardiasis like diarrhea, nausea, vomiting and fatigue. These cysts can survive several days (up to 77 days in water at less than 10°C) and are highly resistant to chlorine disinfection, however they get inactivated when subjected to temperatures of 54°C and above for 5 minutes.

Another important protozoan, the cryptosporidium species, also causes diarrhea. Specifically, *C. parvum* is the major species causing the disease. Human beings are the reservoir for these infectious protozoans and one infected human can excrete 10^9 oocytes a day. *C. parvum* oocysts are 4 to 6 μm in size and spherical in shape. Similar to *Giardia* cysts, *C. parvum* oocysts can survive for several months in water at 4°C and are highly resistant to chlorine.

(d)   Helminths (Eukaryotic)

Helminths are intestinal worms that do not multiply in the human host. For example, hookworms live in the soil and can infect humans by penetrating their skin. With a heavy worm infection, the symptoms are anemia, digestive disorder and abdominal pain. The guinea worm is 0.5–25 mm long, and its eggs may get transmitted through contaminated drinking water supplies. These worms lead to a condition called dracunculiasis and the worms emerge from blisters in a few weeks. Normally, the wound heals rapidly without treatment. Sometimes, the wound may become infected and affect joints, causing significant disability.

6.1.1.2   Inorganic Contaminants

Suspended solids are small particles suspended in water causing cloudiness. Suspended solids present in water are a good indicator of pathogens, since bacteria like to stick to particles in water. Turbidity is a common measure of quality of water. It is an optical property of water and quantifies how clean the water is. The turbidity of water is measured in nephelometric turbidity units (NTUs). The higher the NTU, the more the turbidity in water. Some contaminants can cause the water to have an unpleasant odor, color and taste.

Total dissolved solids (TDS) is the sum of all dissolved chemicals present in water. Most TDS (about 99%) in natural water is contributed by six major ions: calcium, magnesium, sodium, bicarbonate, chloride and sulphate. The balance of TDS is contributed by a number of contaminants like iron, manganese, potassium,

TABLE 6.1
Typical range of total dissolved solids (ppm) in water

Water	TDS (ppm)
Rain water	1–50
Surface water in hills	10–100
Surface water in plains	100–1,000
Groundwater	500–5,000
Groundwater polluted by tannery	10,000–20,000
Sea water	20,000–40,000
Bottled mineral water	100
Permissible limit of drinking water as per WHO	500
Maximum allowable limit for drinking water as per WHO	2,000

ammonia, nitrite, nitrate, fluoride, phosphate and trace metals. There are several kinds of inorganic contaminants. Some of the important ones which affect health adversely are chromium, mercury, arsenic and selenium. Organic contaminants that contaminate drinking water are chemicals such as pesticides.

The quality of water may be different in different places depending on its source. Groundwater generally has low turbidity and high TDS, whereas surface water generally has high turbidity and low TDS. Indicative TDS in water collected from different sources is given in Table 6.1.

Different types of chemical substances are present in water. Iron is present in most water. In deep-well water, it may be present as the soluble, colorless ferrous bicarbonate. Water may look clear and colorless initially, but after some time in contact with air, reddish-brown precipitate of ferric hydroxide may be formed. In some areas, groundwater may have high fluoride content. If these substances are present beyond a particular level, human health is adversely affected. Some health problems like fluorosis, methemoglobinemia (blue baby disease) etc., are associated with chemical parameters. Guidelines are prescribed and standards exist to indicate the desirable and permissible limits of water quality. Table 6.2 gives the WHO guideline values for drinking water and the significance/health effects of chemical parameters.

6.1.2 MONITORING

Measurement and monitoring of the chemical substances and contaminants present in water are important. Contaminants, such as pathogens, are difficult to monitor in the field. Most of the time, water quality is measured in terms of indicator organism, which can be easily tested. The most common ones used are total coliforms, fecal coliforms and *E. coli*. Turbidity is another common measure, as bacteria tend to stick to suspended particles. It is a good indicator of bacteria and is easily measured. WHO has issued guidelines for each of the indicator organisms (Table 6.3). Table 6.4 gives possible inferences from testing for indicator organism *E. coli*.

TABLE 6.2
WHO 1993 Guidelines for Drinking Water and Significance/Health Effects of Some of the Chemical Parameters[3]

Parameters	WHO Guideline Values	Significance/Health Effects
Color (TCU)	15	Consumer acceptance decreases
Turbidity (NTU)	5	Consumer acceptance decreases
pH	<8	Bitter taste, corrosion, aquatic life affected
TDS (mg/L)	1000	Undesirable taste, gastrointestinal irritation, corrosion
Sodium (mg/L)	200	Nausea, convulsions, muscle rigidity, heart-related issues
Chloride (mg/L)	250	Taste affected, corrosive
Iron (mg/L)	0.3	Poor taste, color, turbidity
Fluoride (mg/L)	1.5	Dental and skeletal fluorosis
Arsenic (mg/L)	0.01	Toxic, central nervous system affected, carcinogenic
Chromium (mg/L)	0.05	Carcinogenic, respiratory problems, skin complaints
Lead (mg/L)	0.01	Tiredness, abdominal discomfort, anemia, damage to kidneys
Cadmium (mg/L)	0.003	Toxic, cardio-vascular system affected, hypertension
Mercury (mg/L)	0.001	Highly toxic, neurological issues
Pesticide (mg/L)	0.001	Affects central nervous system
Gross alpha (Bq/L)	0.1	Radiological risk to health
Gross beta (Bq/L)	1.0	Radiological risk to health

TABLE 6.3
WHO Guidelines for Drinking Water with Respect to Coliforms

Indicator	WHO Guidelines Per 100 mL
Total Coliforms	10
Fecal Coliforms	0
E. coli	0

The following sampling procedure is normally followed for bacteriological examination.

- The water is collected in a sterilized glass bottle (250 ml capacity).
- The sample should reach the laboratory within 6 hours from the time of collection. However, if preserved in an ice box, the sample can be delivered within 24 hours.
- The sample collected should be labeled properly.

TABLE 6.4
Inferences from *E. coli* Count

Number of *E. coli* Present	Inference
0–10	Reasonable quality
10–100	Polluted
100–1000	Dangerous
>1000	Very dangerous

Common facilities for monitoring the quality of the water include testing facilities for physical and chemical examination. Physical examination involves monitoring of color, odor, turbidity, conductivity, total dissolved solids and suspended solids. Chemical examination involves monitoring of pH, sodium, iron, manganese, chloride, fluoride, chromium, arsenic, residual chlorine etc. Water for chemical examination is collected in a clean (2-liter capacity) container. The container is washed with the water to be sampled.

- Before sampling, the source is adequately flushed.
- The water is filled completely without leaving any air space.
- Testing is carried out within 24 hours from the date and time of collection.

Source particulars such as location of sampling, type of source, date of collection, pollution to source if any, as well as tests to be conducted and the purpose of testing are noted for each sample.

The water supplied to big cities is usually derived from surface-water sources like lakes, rivers, dams and reservoirs or from groundwater (bore-well) sources. Water from surface-water sources may be turbid and have high bacterial contamination.

6.1.3 Conventional Water Purification

Conventional water purification technology normally consists of physical and chemical treatments. Physical treatment implies conventional filtration techniques including sedimentation and coagulation to remove suspended solids and turbidity from the water. Chemical treatment of water is carried out for the removal of microorganisms and contaminants.

6.1.3.1 Sedimentation and Coagulation

Sedimentation is based on allowing the raw water to remain undisturbed in a container for an extended period of time, which allows suspended solids to settle down.

Coagulation is a process in which suspended solids are made to coalesce to form bigger particles which settle down faster.

Generally, coagulants are first added to the raw water and rapidly mixed for a short time to ease their diffusion in the water. The water is subsequently gently mixed for an extended period so as to increase the rate of contact between the microscopic particles, known as colloids, suspended in the water and the larger solid particles that collect on them. Once agglomerates have reached sufficient size, sedimentation becomes a rapid process. The sedimentation process requires a large clean container in which water is left to settle when coagulants are added. The treated water can be removed from the container's surface without disturbing the sediment at its base.

Size of colloids varies between 10^{-7} and 10^{-9} m. Colloids are negatively charged, attracting positively charged ions from the surrounding water to their surface. The positive charge on the outer surfaces of individual colloids causes them to repel each other. Due to their small size, these fine particles do not easily settle down through sedimentation alone. Adding a coagulant to the water alters the electrostatic situation of the colloids by agglomerating. Once smaller insoluble agglomerates of colloids have formed, they start coalescing with larger suspended solids such as fecal contaminants.

a) Inorganic coagulants

Alum is one of the inorganic coagulants and carries a net positive charge. Its diffusion in water causes the positive outer shells of surrounding colloids to be compressed against their neutral cores, thereby neutralizing their apparent charge. The colloids begin to agglomerate in small amounts due to simple gravitational interaction.

The following chemical equations show how alum, $Al_2(SO_4)_3$, reacts with a variety of common colloidal particles found in water, such as in the presence of calcium bicarbonate, $Ca(HCO_3)_2$:

$$Al_2(SO_4)_3 + 3Ca(HCO_3)_2 \leftrightarrow 2Al(OH)_3 + 3CaSO_4 + 6CO_2 \qquad (6.1)$$

The rate at which sedimentation occurs is given by the following equation.

$$V_g = 1/18 \times (\rho_s - \rho)g\,d^2 / \mu \qquad (6.2)$$

where
   V_g: speed with which particles settle to the bottom
   ρ_s : solid particle density
   ρ : density of water
   d : solid particle diameter
   g : gravitational acceleration (9.8 m/s²)
   μ : viscosity of water

By increasing particle diameter and thus the speed of sedimentation, coagulation and flocculation technology is capable of removing water turbidity up to 90% along with corresponding removal of associated microorganisms and pathogens. The amount of time necessary for both the initial rapid and secondary slower mixing stages depends on the specific coagulant in use and the turbidity of the water. With respect to alum, the necessary dose depends on pH and turbidity of raw water prior to treatment. Dosing of alum of 11 mg/L to as high as 30 mg/L may normally be required to reduce turbidity levels to below 5 NTUs.

b)   Organic coagulants

A variety of biological coagulants are used across the world. The prickly pear, a cactus commonly known as *tuna* or *raquette* in Haiti and Chile respectively, is a popular biological coagulant. Other examples of naturally occurring coagulants include potato starch, dried beans, peach seeds, cattails, totora, water hyacinth and duckweed. Use of *Moringa oleifera* seeds as a coagulant has increased significantly over the years. *Moringa oleifera* powder must be prepared fresh before use in order to ensure its effectiveness. Humidity can cause deterioration of the active agent within the seeds.

The mechanism by which biological coagulants function is still under study. In the case of *Moringa oleifera* seed powder, it is dependent upon a long polypeptide protein chain. This polypeptide chain may act in one of two ways. First, it has a net charge opposite to that of the colloid. It is thus attracted and bonds with the chain, allowing larger agglomerates to form. Second, the polypeptide chain is able to form a bridge between two colloidal particles by interacting across both their positive shells and negative cores. The effectiveness of *Moringa oleifera* seed powder depends on a variety of biological and local factors. Use of *moringa* neither alters pH value of raw water nor is *moringa*'s effectiveness affected by pH. Its ability to reduce turbidity depends on the properties of local seeds. No toxic effects have yet been reported with the use of *Moringa oleifera* seed powder for water treatment.

The processing environment used in sedimentation, coagulation and flocculation at the household level is a single large container, tall enough to allow a clear differentiation between clean water on top and residual water on the bottom. The container should be relatively clean and free of debris. Alum is generally used in water treatment plants at community level with sufficient expertise and skills, due to the toxicity of an alum overdose.

6.1.3.2   Chemical Disinfection

Chlorine is a common chemical disinfectant. One of the common methods for generatingchlorine is salt water electrolysis, resulting in production of sodium hypochlorite (NaOCl).

$$H_2O + NaCl = NaOCl + H_2 \tag{6.3}$$

Sodium hypochlorite (5–15% conc), when added in raw water, reacts as per following equation:

$$NaOCl + H_2O = HOCl + NaOH \tag{6.4}$$

The compound HOCl is a weak acid known as hypochlorous acid. When dissolved in water, hypochlorous acid is dissociated into a single proton, H^+, and hypochlorite ion, OCl^-. Generally, their pathogen elimination effectiveness is relatively low.

Another form of chlorine commonly used is solid calcium hypochlorite, i.e., $Ca(OCl)_2$. It is more corrosive than the bleach solution, NaOCl. Special measures are taken when handling it to prevent inhalation or direct skin contact. In order to make the $Ca(OCl)_2$ chlorine useful at the household level, it is dissolved in water similar to that of NaOCl-based bleach. In solid form, assuming no major contact with moisture, $Ca(OCl)_2$ can be kept for a long period of time. Once dissolved, its shelf life is limited. Effectiveness of chlorine disinfection and the dose requirement depend on water turbidity. As water turbidity increases from 1 NTU to 10 NTU, chlorine demand increases eight-fold. While a cut-off point at which chlorine treatment becomes entirely ineffective has not yet been determined, it is recommended that water with a high NTU value should be pretreated by filtration, coagulation or sedimentation prior to chlorine use.

The effectiveness of chlorine in microorganism destruction is commonly given by the breakpoint curve, shown in Figure 6.1.

As shown in Figure 6.1, initially chlorine is consumed by the raw water through reactions with the various microorganisms in the raw water as well as a variety of impurities such as Fe^{2+}, Mn^{2+}, H_2S, etc. In the next stage, the chlorine residual remaining in the water begins to increase as the chlorine reacts with ammonia and the organic matter of microorganisms within the water. In the third stage, the chlorine begins to oxidize chloroamines formed by reactions with ammonia in the previous stage of the curve. This leads to a temporary decrease in chlorine levels. The point at which the third stage transfers to the fourth is known as the breakpoint. The amount of chlorine required to reach this point of the curve is known as the chlorine demand of the water. Any chlorine added after the breakpoint subsequently remains in the water as hypochlorous acid (HOCl) or hypochlorite ion (OCl^-). This remaining chlorine, referred to as the chlorine residual, is left in the water to prevent recontamination, allowing water to be stored for more extended periods of time than would ordinarily be possible, without fear of recontamination. Another factor affecting chlorine treatment is temperature. Higher temperatures lead to a quicker rate of reaction between the chlorine and the various microorganisms suspended in the water.

While both hypochlorous acid and hypochlorite ion have disinfection properties, hypochlorous acid is up to 80 times more effective. The amount of HOCl dissociated

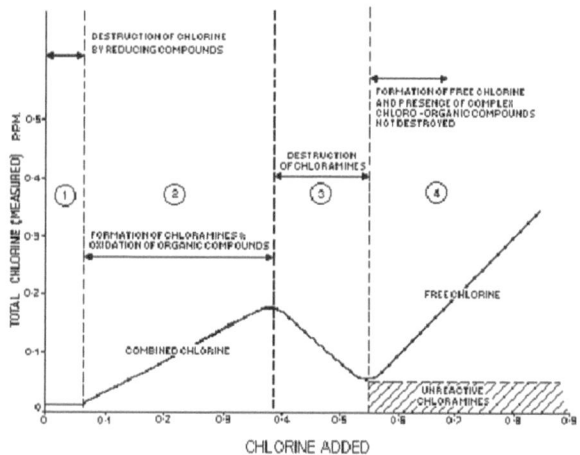

FIGURE 6.1 Chlorination breakpoint curve.

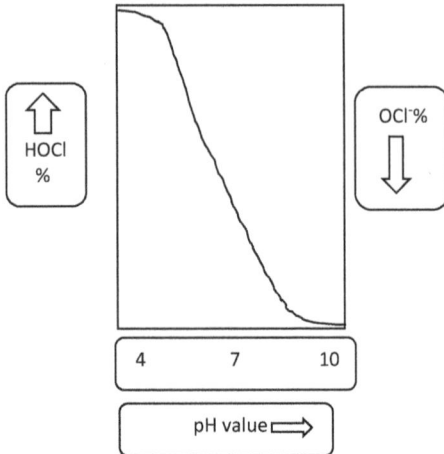

FIGURE 6.2   Effect of pH on HOCl and OCl⁻ distribution.

in the manner outlined above is dependent upon pH of water. Though HOCl is more effective than OCl⁻ in the destruction of pathogens, the increased pH will result in lower chlorine performance, as shown in Figure 6.2.

The proportion of HOCl and OCl⁻ is a function of temperature as well as pH. As temperature increases, the curve shifts gradually to the left. The effectiveness of a chlorine dose is governed by a combination of dose concentration and contact time. This is typically represented by the following equation, known as Chick's Law, in which Nt is the number of microorganisms in the water at time t, and k is a constant term.

$$\left(dN_t/dt\right) = -kN_t \tag{6.5}$$

The effectiveness of chlorine disinfection is proportional to the amount of free available chlorine in the water. One of the common measures of chlorine effectiveness is the Ct measure, a combination of chlorine concentration and contact time, given by Equation (6.6):

$$C_t = C_{xt} \tag{6.6}$$

where C is the concentration of free chlorine residual within the water and t is the elapsed time since the initial dose.

In general, bacterial species exhibit the lowest resistance to chlorine, followed by viruses and protozoa. The ability of specific microorganisms to be inactivated by chlorine disinfection can be represented by organism-specific C_t values. Microorganisms, such as *Gardia lamblia*, with high C_t values exhibit higher resistance to chlorination. C_t values typically range from less than 0.01 mg/L·min to greater than 100mg/L·min. This implies that complete disinfection can take anywhere from less than one minute to over 100 minutes at free chlorine residual levels of 1.0 mg/L. With the exception of, which forms a protective cyst upon contact with chlorine, most protozoans are only fractionally more resistant to chlorine treatment than bacteria and viruses. Table 6.5 shows a number of significant pathogens and their relevant Ct values.

TABLE 6.5
C_t values for a Variety of Pathogens

Bacteria	Cl$_2$ Residual (mg/L)	Temperature (°C)	pH	Time (min)	Reduction (%)
Campylobactor jejuni	0.1	25	8.0	5	99.99
Escherichia coli	0.2	25	7.0	15	99.99
Legionella pneumophila	0.25	21	7.8	60–90	99
Mycobacterium chelonet	0.7	25	7.0	60	99.95
Mycobacterium fortuitum	1.0	–	7.0	30	99.4
Mycobacterium intracellulare	0.15	–	7.0	60	70
Pasteurella turarensis	0.5–1.0	10	7.0	5	99.6–100
Salmonella typhi	0.5	20	–	6	Two orders of magnitude
Shigella dysenteriae	0.05	20–29	7.0	10	99.6–100
Staphylococcus aureus	0.8	25	7.2	0.5	100
Vibrio cholera (smooth strain)	1.0	20	7.0	<1	100
Vibrio cholera (rugose strain)	2.0	20	7.0	30	>5 orders of magnitude
Yersinia enterocolitica	1.0	20	7.0	30	92
Virus	**Cl2 residual (mg/L)**	**Temperature (°C)**	**pH**	**Time (min)**	**Reduction (%)**
Adenovirus	0.2	25	8.8	0.7	99.8
Coxsackie	0.16–0.18	27–29	7.0	3.8	99.6
Hepatitis A	0.42	25	6.0	1	99.99
Parvovirus	0.2	20	7.0	3.2	99
Poliovirus	0.5–1.0	25	7.4	30	100
Rotavirus	0.5–1.0	25	7.4	30	100
Protozoa	**Cl2 residual (mg/L)**	**Temperature (°C)**	**pH**	**Time (min)**	**Reduction (%)**
Cryptosporidium parvum	80	25	7.0	90	90
Entamoeba histolytica	1.0	22–25	7.0	50	100
Giardia lambia	1.5	25	7.0	10	100

Chemical disinfection is one of the most effective forms of treatment for the removal of a wide variety of protozoa, bacteriological and viral pathogens. The use of the chlorine residual makes it possible for treated water to be stored without significant risk of recontamination. It is one of the commonly used forms of water purification in many areas. Use of $Ca(OCl)_2$ often produces large volumes of hazardous concentrated waste. Additionally, the dry storage required for $Ca(OCl)_2$ may prove difficult to maintain in tropical regions where moisture is present. Overdose of chlorine results in unpleasant taste and odor.

By-products of chlorine treatment are carcinogens known as trihalomethanes (THMs). Although chronic exposure to THMs may lead to cancer during later life, the prevalence of debilitating sickness caused by waterborne pathogens has led the WHO to issue this statement: "Where local circumstances require, a choice be made between meeting microbiological guidelines or guidelines for disinfection by-products such as chloroform, the microbiological quality must always take precedence. Efficient disinfection must never be compromised."

Ozone, an oxidizing agent, is also a quite effective disinfectant. The extra atom of oxygen in ozone destroys odor, bacteria or viruses by oxidation. During this process, the extra atom of oxygen is consumed. Ultraviolet (UV) radiation is also used for disinfection. When exposed to sunlight, germs are killed and bacteria and fungi are prevented from spreading. This natural disinfection process can be utilized effectively by applying UV radiation in a controlled manner.

6.1.3.3 Biosand Filtration

Biosand effectively removes suspended impurities, turbidity, certain organics and pathogens from raw water. Figure 6.3 illustrates biosand filtration at household level. Water is manually poured into the filter. The water travels through the diffuser plate, then the various sand media, before finally flowing into the piping and out of the filter, where it is collected for use.

A biosand filtration unit consists of a casing, which is an outer container that keeps the components of the filter together as a single unit. A lid is used for keeping dirt or other unwanted impurities out of the filter. The diffuser plate is placed above the sandbed, which dissipates the initial force of the water. Sediments, cysts and worms are trapped in the spaces between the sandgrains. Certain organic components are adsorbed as well. Water collects in the pipe at the bottom of the filter. The water then travels through the pipe and out of the filter for use. The water supplied to the filter is raw water obtained from rainwater, deep wells, shallow wells, rivers, lakes, reservoirs or surface water. The biosand filter can purify raw water with turbidity up to 100 NTU.

The removal of pathogens in the biosand filter is achieved by a combination of biological and mechanical processes: predation, natural death, adsorption and mechanical trapping. Water contaminated with organic material is poured into the filter, which is trapped on the surface of the fine sand layer and forms the *schmutzdecke* biological layer. Microorganisms colonize this layer over a period of 1–3 weeks. Once established, they consume pathogens present in water. As the environment is not favorable to pathogens, natural death occurs. In the adsorption process, viruses and organic contaminants get adsorbed to sandgrains. Sediments, cysts and worms get trapped in between the sand grains.

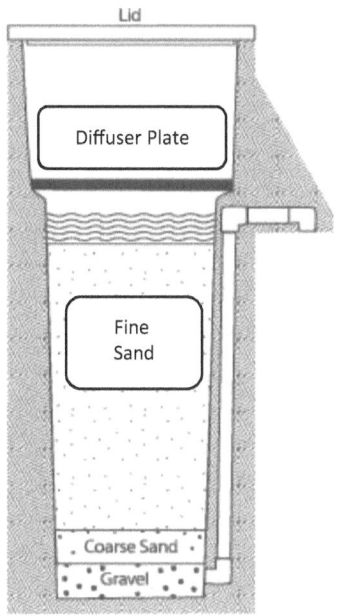

FIGURE 6.3 Biosand filtration.

A 50 mm layer of water is maintained on top of the fine sand layer. The water level is shallow enough to allow oxygen to diffuse in the *schmutzdecke* for biological development, yet deep enough to keep the sand from drying out. It is desirable that the water poured in the filter should be collected from a consistent source to ensure a biological balance.

The biosand filter can produce up to 60 liters of water in an hour. The exact flow rate is controlled by the size of the sand media. There is a trade-off between flow rate and efficiency of the biosand filter. When the rate is excessive, efficiency is reduced, whereas if rate is insufficient, there will be less production of purified water (Figure 6.4).

The biosand filter is quite effective and can effectively remove disease-causing organisms from water. The minimum pause period between addition of water is one 1 hour and the maximum is 48 hours. Removal capabilities of biosand filter are given in Table 6.6.

The main ingredients required in a biosand filter are:

a) Cement for the casing
b) Wood for lid and diffuser
c) Fine sand
d) Coarse sand
e) Gravel
f) Plastic tubing

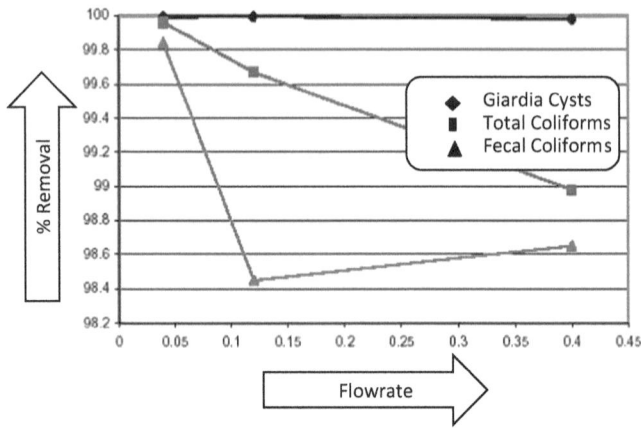

FIGURE 6.4 Effectiveness of Biosand filter.

TABLE 6.6
Biosand Filter Removal Capabilities

Type of Contaminant	Removal Percentage (%)
Turbidity	100
Protozoa	100
Helminthes	100
Fecal coliform	90
Zinc, copper, cadmium and lead	95–99
Iron and manganese	<67
Arsenic	<47
Suspended sediments	100

Source: CAWST (Centre for Affordable Water & Sanitation Technology)

Cement is readily available in most parts. A small amount of wood is required for constructing the lid and the diffuser plate. The wood can be replaced by other materials if they are more suitable or available locally. The sand media is a key ingredient of the filter.

The advantages of the biosand filters are manifold. The filter is durable and lasts for several years. Since production of the filter does not require a large or specialized facility, manufacturing can take place locally in the community itself. Maintenance is simple. The unit can be locally produced and maintained in rural and remote areas. However, the performance of the biosand filter for removal of microorganism and turbidity needs to be assessed thoroughly before implementation of the technology. At times, additional treatment may be necessary to ensure that all bacteria in water are inactivated or killed.

6.1.3.4 Ceramic Filtration

Ceramic filters are also known as white candle filters. The red clay (terracotta) candle and disc are also developed, which are quite useful in rural areas for reducing turbidity and microorganisms in water.

a) White Ceramic Candle Filter

A white ceramic candle filter consists of two chambers and one or multiple ceramic filter elements, shaped like a candle. The candle is usually made of white kaolin clay, quartz, etc. The mixture of clays is powdered and mixed with combustibles such as wood charcoal powder. It is mixed with water to make thick dough or clay water slurry. The dough is molded into a candle shape under pressure. Alternatively, the clay water slurry can be casted in a plaster of Paris mold to form the candle. Then, the candle is dried and subjected to high temperature in a furnace. It is a hollow cylindrical shape of which one end is closed and the other open. The open end is fixed in a metallic/ plastic threaded holder through white cement. The end of the holder is fitted at the bottom of the container to filter raw water. The filtered water is collected in the lower chamber for household use.

A large number of pores are created in the candle due to burning of combustibles. The pores are interconnected, leading to capillary paths from the outer surface to the inner bore of the candle. Generally, the size of the pores is in the range of 0.3–6 μm depending on combustibles, construction materials and temperature of burning. Ideally, a filter needs to have a pore size of 1 μm or less. Hence, the average rate of removal of microorganisms by the white candle is up to 85%. It filters about 0.3–0.5 liters of water in an hour from a candle. The pores or the capillary paths generally get clogged during the course of its use. The bigger suspended particles, microorganisms etc., clog the capillary paths during filtration. Therefore, the outer surface of the candle needs to be scrubbed for removal of sediments from the candles. The rate of filtration reduces with duration of its use. Since the pores are interconnected, it requires the candle to be replaced when its core gets clogged.

b) Terracotta Water Filter

The terracotta water filter is another filtration and treatment device for water purification. It consists of a burnt red clay porous medium produced from the mixture of red clay (silt clay). Terafil, an improved terracotta water filter, is prepared from a mixture of ordinary pottery clay (red clay), river sand and wood sawdust (combustibles) which are widely available in rural and remote areas.[1] The mixture is sintered in a furnace to the desired shape. The water filter has two pots, placed one over another (Figure 6.5).

The top pot of the filter has a small opening at its bottom, over which the Terafil is fixed by cement or by other means. The filtered purified water is collected in the lower pot. The device removes sediment, suspended particles, microbes, iron, heavy metals, color, odor etc., effectively from the turbid raw water. The filtration rate is rejuvenated by periodic scrubbing and cleaning the surface. The design of the domestic filter can be scaled up for community use with gravity and pressure flow to get a high rate of filtration, 60–1500 liters in an hour (Figure 6.6).

FIGURE 6.5   Terafil water filter.

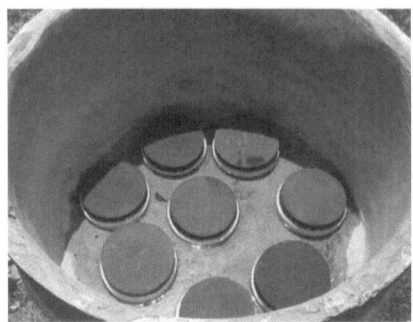 

FIGURE 6.6   Terafil water filter for community use.

During the sintering process, the wood particles are burnt; clay particles are sintered around the sand particles, leaving large circular pores in between. The pores are separated by thin clay membranes of 50–100 µm thickness. Sintering of Terafil plays an important role in obtaining proper thickness of the membrane as well as the shape of the pores and Terafil strength. The large pore works like a micro-reservoir for water inside the Terafil. Water travels from pore to pore in the clay membrane inside the Terafil during filtration The thickness of the clay membrane determines the flow resistance. Rate of flow is greater in the thin membrane as it reduces the flow resistance. The sintered clay membrane is a composite of multiple ultra-fine sintered clay particles (layers). The ultra-fine clay particles are separated from each other by capillary or ultra-fine openings. The capillary openings of a membrane bridge a set of large pores created due to burning of sawdust. The diameter of the interface or the capillary is in sub-micron size, 0.2–0.3 µm. Generally, the harmful pathogens are between 1 and 50 µm in size, which is more than the diameter of the capillaries. So the travel of microbes, suspended particles, etc., along with the water is restricted. The Terafil is activated during the sintering process and the clay is negatively charged. Therefore, soluble iron and some heavy metals of raw water are also removed by adsorption. The filtered water flows from one micro-reservoir to another inside the Terafil through the capillary openings of the clay membrane due to the semipermeable nature of clay. The rate of flow of the filtered water through the capillaries also depends on the head of the water column.

Removal of suspended particles, iron, microorganisms, etc. occurs at the top surface of the Terafil, that is, the surface exposed to raw water. Over time the contaminants clog the top surface during use and reduce the flow rate. The top surface of the Terafil must be scrubbed with a nylon brush or similar material to remove the sediment, open new pores and rejuvenate the flow rate. Studies have been conducted on important parameters such as pH, turbidity, iron, coliform bacteria and rate of filtration. The pH value of filtered water is maintained within 6.5–8.5 as per drinking water standard. High-turbidity raw water (up to 500 NTU) and high-concentration coliform organisms (1100 MPN/100 mL) can be treated and purified to turbidity less than 1 NTU and bacterial count less than 7 MPN/100 mL, suitable for drinking as per WHO standard. It also reduces more than 90% soluble iron from raw water. However, it should be ensured that there is no leakage through the joint between the container and Terafil. Since the filters can be produced locally, there is a greater chance of sustainability in the long term particularly in rural and remote areas. A Terafil filter having containers made of stainless steel or food-grade plastic appear promising. The Terafil water filter can remove up to 90% soluble iron, 99% turbidity and 95% microorganism.

6.1.3.5 Rice-Husk Concrete Filtration

Concrete blocks made of rice-husk ash, pebbles and cement can be used as filtration device for the purification of raw water contaminated with bacteria and turbidity. The rice-husk ash, pebbles and cement in proper proportion are mixed with water, and the concrete mixer is rammed into a mold under a nylon mesh. After ramming, the top end of the bed is covered with another nylon mesh. The nylon mesh stops the fine ash coming from the filter element. The settling characteristic of cement imparts the necessary strength to the element. The output parameters, like bacterial trapping, turbidity, pH and filtration rate, are monitored. The trapping efficiency for bacteria and turbidity is up to 99% under laboratory conditions. However, there is considerable variation from filter to filter when produced in large numbers. Variables like the type of ash, the type of pebbles, the mixing process, the ramming process, testing procedures, etc. influence the output results. It is the identification of the important variables between these and their control that enables prediction of the filter characteristics when fabricated in larger numbers. The tolerance for the trapping characteristics for bacteria and turbidity is about 95±4%. The performance of the concrete filter is satisfactory when maximum input turbidity of raw water is up to 50 NTU to achieve output turbidity of less than 1 NTU. Similarly, when the input bacteria concentration increases, the output water quality also deteriorates. This method has limitations for purifying raw water of very high turbidity.

6.1.3.6 Colloidal Silver-Based Filtration

Colloidal silver is formed by running a positive electrical current through bars of pure silver suspended in water. The colloidal silver particles have a size ranging from 0.015 to 0.005 μm. It has a positive ionic charge and is used in many applications as an antibacterial agent.

The colloidal silver is coated on a ceramic candle or disc to kill residual microorganisms during filtration of raw water. The colloidal silver acts as an antibacterial agent or germicide. Silver colloids attack specific enzyme pathogenic bacteria and fungi. By attacking the food source rather than the bacteria, the silver prevents micro-organisms developing a resistance to it. Colloidal silver attacks other pathogens with its electric charge, causing their internal protoplast to collapse. Other effects include making bacteria unable to reproduce and killing parasites in their egg stage. However, colloidal silver cannot penetrate the core and always sticks over the surface ceramic candles. Since the surface of the candle is to be cleaned from time to time for removal of sediments, repeated colloidal silver treatment is advisable for achieving a guaran-teed 100% removal of microorganisms.

6.1.3.7   Ion Exchange as Water Softener

Ion-exchange systems consist of beds of synthetic resin for selectively adsorbing certain cations or anions and replacing them with counter-ions. The process of ion exchange continues till all available spaces are filled up with ions. The ion-exchange device is regenerated by suitable regenerating chemicals. The water softener is one of the common ion exchangers, normally used to remove hardness from raw water. It removes calcium and magnesium ions from hard water, replacing them with other positively charged ions, such as sodium. When hard water is passed through a sodium cation exchanger, the calcium and magnesium ions are adsorbed and held by the cation exchanger, which simultaneously gives up an equivalent amount of sodium in exchange. When the ability of a cation exchanger bed to produce soft water is exhausted, the softening unit is temporarily put out of service and back-washed. Cation exchangers have a stronger absorptive power for bivalent cations than for monovalent cations. When hard water, which is a dilute solution of cal-cium, magnesium and sodium salts, comes in contact with a sodium cation exchanger, the bivalent cations of calcium and magnesium are adsorbed by the cation exchanger, which simultaneously releases an equivalent of monovalent sodium cation in exchange. Hard water contains ferrous and manganese bicarbon-ate; the iron and manganese can be removed by a sodium cation exchanger simulta-neously with the removal of hardness. Sodium chloride is widely used for regeneration of the exhausted cation exchanger bed. Sodium nitrate is also an excel-lent regenerant, but relatively more expensive. Potassium chloride, potassium nitrate and sodium sulphate can also be used for regeneration. Sodium cation exchanger water softeners are made in both the pressure type and gravity type. The regeneration consists of three steps: (i) backwashing, (ii) salting or brining, (iii) rins-ing. Backwashing occurs when a strong flow of water moves upward through the cation exchanger bed to expand, cleanse and restore. In sodium cation exchange, the water is completely softened; however, the TDS content is not decreased and the effluent contains the same amounts of anions as the raw water as well as sodium cations equivalent in amount to those originally present. Calcium and magnesium salts present in raw water are simply replaced by equivalent amounts of correspond-ing sodium salts. In hydrogen cation exchange, hydrogen ions are exchanged for calcium, magnesium and sodium ions. Thus, calcium, magnesium and sodium ions are removed from water. The theoretical carbonic acid formed from the bicarbonate breaks down into water and carbon dioxide. Similarly, sulfuric acid and hydraulic acid, corresponding to the amount of sulphate and chloride present in the raw water, are left in the effluent. In the hydrogen cation-exchange process, regeneration is done by sulfuric acid or hydrochloric acid.

6.1.4 Recommendations

Several types of water purification device have been developed to obtain clean and safe drinking water. Many are established in the market as commercially viable products. Selection of appropriate technology is an important aspect. It is necessary to identify cost-effective, energy-efficient and environment-friendly technology on a case-by-case basis. It is recommended that development efforts are directed through technological innovations to improve yield, reduce energy requirement and ensure long operating life. Existing technologies need to be improved and emerging technologies have to be field tested and evaluated from various angles for commercial-environmental suitability and scale-up.

Further efforts need to be directed to improve performance, keeping in mind the end use, whether mass production is by village artisans or industry, the robustness, durability and maintenance of the device, and its sustainability for society.

A performance comparison of water purification devices is given in Table 6.7.

TABLE 6.7
Water Purification Devices Performance Comparison

S. No.	Name of Technology	Reduction of Contaminants (%)	Remarks
1	Inorganic coagulants (alum)	Reduction of (a) turbidity (90%), (b) microorganisms (90%)	Maintained by user
2	Organic coagulants (moringa seed)	Reduction of (a) turbidity (90%), (b) microorganisms (90%)	Maintained by user
3	Chlorine treatment, (bleaching powder)	Reduction of microorganisms (100%)	Maintained by user
4	Biosand filter	Reduction of (a) turbidity (99%), (b) coliform (70%) (c) iron (70%) (d) Heavy metals (95%)	Maintained by user
5	White ceramic candle	Reduction of (a) turbidity (90–95%), (b) microorganisms (85%)	Maintained by user, works well with low- turbid water
6	TERAFIL Terracotta filter disc	Reduction of (a) turbidity (99.9%), (b) microorganisms (95–100%) (c) iron (90–95%) (d) color (99%)	Maintained by user
7	Rice-husk ash candle filter	Reduction of (a) turbidity (98%), (b) microorganisms (98%) in laboratory conditions	Maintained by user
8	Colloidal silver-based unit	Reduction of (a) microorganisms (100%) with the help of candle/ Terafil filter	Maintained by user

6.2   INDUSTRIAL WATER TREATMENT

The quantity and quality of water required by a particular industry depends on its type. It also depends on the purpose for which it will be used in the industry. Seawater, for example with chlorination, is suitable for cooling applications and yet absolutely unsuitable for boilers and several other applications. In some cases, ground or river water itself may be suitable and there is no need for treatment. In other cases, a single treatment, such as softening, may be necessary. For certain uses, however, the quality of water required may be so high that complete removal of all the contaminants is required.

6.2.1   Industrial Water Uses

The water used in industrial plant may be classified into the following major categories

- Process water
- Boiler feed water
- Cooling water
- General-purpose water

The quality requirement of the water varies according to the application. It may differ even for the same application, for example, the quality of boiler feed water required for a low-pressure boiler is not the same as for a high-pressure boiler. Similarly, the quality requirement for cooling water or process water in a particular case may not necessarily be the same as the quality requirement for cooling water or process water in another case.

6.2.1.1   Process Water

The quality of water required for different processes varies over a wide range. In many cases, the quality of water has an impact on the quality of the finished product and influences the economics of the processing. Some processes may not require any treatment or only minimal treatment of raw water for use, while some other processes may need only a reduction in bicarbonate hardness. On the other hand, quite a few industrial processes may require very pure high-quality water, such as distilled water or demineralized water free from impurities.

Process plants are generally large users of water. Several of them treat and reuse part of their wastewater for certain applications in the plant. In the case of surface water containing turbidity, color and other impurities, preliminary treatment for removal or reduction of the impurities to the specified level is commonly practiced. Subsequent treatment depends on its end use. Some of the process plants use steam as process steam for requirements such as a reactant. In industries operating their steam boilers at moderate pressure, sodium cation exchange is widely used to soften the boiler feed water. In some cases, it is preceded by a cold lime process that lowers the alkalinity and total solids in the final effluent. Hydrogen cation exchange, with the acid effluent neutralized by caustic soda or with the effluent from sodium cation exchange, is also used for softening the boiler feed water and generating process steam. In plants operating at higher pressures, the hot water softening and silica removal process is used. For higher pressures, the ion-exchange demineralization process, involving cation and anion exchange followed by silica removal, is widely used.

6.2.1.2 Boiler Feed Water

Apart from producing process steam, boiler feed water is also required for generating steam for different applications such as heating and utilities. In the case of boiler feed water for low-pressure boilers, the removal of hardness in water by the sodium cation-exchange (zeolite) process is sufficient. For boilers operating at medium pressure, a reduction of TDS and alkalinity, as well as removal of hardness, is required by the two-stage cold lime and sodium cation-exchange process or hydrogen cation exchange followed by neutralization with caustic soda. For high-pressure boilers, removal of hardness and reduction of TDS, alkalinity and silica content is needed. This is achieved by two-stage processes such as water softener and sodium cation exchange. For boilers operating at very high pressure, a complete removal of all contaminants, impurities and dissolved solids is important, which is generally done by ion-exchange demineralization involving cation exchange, anion exchange and mixed bed ion-exchange resins. The degree of deaeration required for all boilers except low-pressure boilers is so high that it requires practically complete removal of all dissolved gases, which is achieved by chemical means by adding appropriate chemicals or a mechanical deaerator operating at atmospheric pressure or sub-atmospheric pressure.

6.2.1.3 Cooling Water

The amount of cooling water required in industries varies from a small fraction of the total water consumption to over 90% of it. Treatment of cooling water depends on the composition of the raw water. It also depends on the type of cooling, which could be once through or recirculation. In the once-through case, cooling water is used once and then discharged as effluent. In such cases, either (i) no treatment, (ii) chlorination, or (iii) only acid treatment of water is sufficient. If the cooling water is recirculated, treatment for prevention of scale and organic growth is usually required to make it suitable for both cooling and subsequent uses. The treatment is carried out using the cold lime process, sodium cation exchange or the two-stage cold lime and sodium cation-exchange process. The reduction of the calcium bicarbonate hardness in water is achieved by adding acid. At times, treatment by cation exchange or demineralization by ion exchange is also done. Chlorination of the recirculated water may also be carried out to avoid biofouling. Since magnesium carbonate is about five or six times as soluble as calcium carbonate, and since magnesium content in water is usually much lower than the calcium content, magnesium plays a very minor role in scale formation. Magnesium bicarbonate is also much more soluble than calcium bicarbonate. However, care should be taken to blow down enough water to keep concentration of the scale-causing constituents well within limits when water is recirculated.

6.2.1.4 General-Purpose Water

In addition to boiler feed and cooling applications, water of appropriate quality is required for general purposes in industry, such as toilets, showers, general flushing, emergency situations, etc. It may need minor treatment such as hardness removal by sodium cation exchange. Water used for flushing purposes seldom requires treatment.

6.2.2 WATER TREATMENT IN INDUSTRY

Treatment of raw water is required to obtain the desired quality for industry. Treatment methodology depends on a number of factors, including composition of the raw water and the quality of the treated water required by the industry.[2] The Langelier saturation index (LSI) helps to approximate the degree of treatment required. It is also known as the calcium carbonate saturation index and helps in predicting the scale-forming tendency of water. LSI is an equilibrium model derived from the saturation concept and provides an indicator of the degree of saturation of water with respect to calcium carbonate. In order to calculate the LSI, it is necessary to have (a) methyl orange alkalinity, (b) the calcium hardness, (c) the total solids, (d) the pH value and (e) the temperature to which water will be raised. LSI is basically a methodology to determine if water is corrosive (negative LSI) or scale forming (positive LSI). LSI between −0.3 and +0.3 is the widely accepted range, though 0.0 is the ideal.

Surface water is relatively more turbid than groundwater. Turbid water may lead to clogging and deposit formation. At times, raw water may contain some color. In general, the removal of turbidity and color is accomplished by coagulation, settling and filtration. As the nature of the organic color in water differs, some high-color waters require coagulation at lower pH values followed by filtration.

For the removal of objectionable tastes and odors, activated carbon is used, sometimes preceded by aeration. Activated carbon in powdered form may be applied in the settling tank, or in other cases, the tastes and odors are removed by passing the water through an activated carbon filter which contains a bed of granulated activated carbon.

Iron and manganese content in water are objectionable because of their staining properties and catalytic effects. Iron and manganese bacteria, popularly known as crenothrix, are quite troublesome, as they form luxuriant masses. These masses frequently break loose and block the passage for cooling water. Bicarbonates of iron and manganese, which are soluble in water, can be removed by aeration, settling and filtration. If the pH value is more than 7.0, iron oxidizes rapidly to insoluble hydroxide, but manganese requires a higher pH value. When hardness reduction is also required, aeration before lime soda process leads to excellent removal of both iron and manganese. If complete softening by either sodium cation or hydrogen cation exchange is required, the soluble iron or manganese is removed simultaneously with the hardness by either of these processes. If iron or manganese is not more than 1 ppm, manganese zeolite filtration can be used for removal. Iron and manganese in acidic water may be removed by aeration, neutralization, settling and filtration.

Hardness is objectionable in many operations, including wet processing operations, and requires softening by sodium cation exchange. Most sulfur waters contain a small amount of hydrogen sulfide and if pH value is not too high, they are usually aerated and then chlorinated to oxidize any residuals.

The following water treatment processes are generally used in industry for the removal of suspended solids, turbidity, color and organic matter from water:

- Sedimentation
- Coagulation
- Settling
- Filtration

The above processes are usually combined as (i) coagulation and settling, (ii) coagulation and filtration, (iii) coagulation, settling and filtration and (iv) sedimentation, coagulation, settling and filtration.

Sedimentation, as mentioned earlier, is the process in which suspended matter is removed without the aid of an added coagulant, by slowing the flow of the water and increasing retention time. The degree of removal of suspended matter depends on the size and nature of the suspended particles in water, temperature of the water and retention period. As these particles may vary in size over a wide range from coarse granules to colloidal dimensions, usually the removal of suspended matter is partial in sedimentation. Much of the turbidity and most of the color are composed of such small particles that it is necessary to treat the water before filtration so that these particles agglomerate together into lumps large enough to be filtered out. This is done using coagulants. The process is known as coagulation and gives crystal-clear water. The coagulants used are compounds of aluminum or iron, usually sulphates. These are acidic in nature and react with the natural or added alkalinity of water to produce sulphates of sodium, calcium or magnesium, and gelatinous precipitate assumed to be hydroxide. Most of the colloidal particles of turbidity and color carry a negative electrical charge, whereas metallic hydroxide carries a positive charge. The dosages of coagulant required and the optimum pH value for coagulation vary with quality of water. In general, the dose rate varies from 1 to 5 ppm. The favorable pH for aluminum coagulants usually ranges from 5.5 to 6.8. The favorable pH for iron coagulants ranges from 3.5 to 5.5 and above 9.0. Color is usually best removed at pH values below 6.5. Controlled agitation helps in good floc formation which settles easily. The dose rate of coagulants and optimum pH value are determined by jar test. The time required for good floc formation is measured. The time required for settling is noted after stopping the agitation, and is expressed as percentages settled over various periods such as 5, 10 and 15 minutes. The jar test helps in indicating the approximate dose rate and optimum pH values. Aluminum sulphate, ferric sulphate, ferrous sulphate, alum and sodium aluminate are commonly used as coagulants in industries. Certain materials, known as coagulant aids or flocculation aids, are used in conjunction with coagulant to aid the removal of suspension solids, color, organic matter etc., by formation of rapidly settling floc. Certain clays, activated silica and polyelectrolytes are widely used coagulant aids.

The bulk of the floc with its entrained loads of impurities is removed by settling. Conventional basins or the sludge-blanket type of coagulation and settling equipment are used extensively for coagulation and settling. For water which requires clarification as well as softening, the two processes can be carried out simultaneously in the sludge-blanket type of equipment, where all the coagulated water is filtered upwardly through a suspended sludge blanket. This intimate contact with previously formed flocs enables its adsorbent powers to be used to the fullest extent. The equipment is made in both vertical and horizontal designs. The raw water and coagulant first flow down through one compartment with increasing cross-sectional area from top to bottom equipped with mechanical agitators. On emerging from bottom of the first compartment, the flow of water is reversed from bottom to top through a suspended blanket of previously precipitated sludge in a second compartment with increasing cross-sectional area from bottom to top so that the velocity of water is too low to lift the floc particles. The sludge blanket is kept within certain limits by bleeding off sludge as new sludge is being formed. The outlet water is relatively clear and good for use as cooling water and for other industrial purposes which do not require perfectly clear water.

In case water is to be further treated, as for boiler feed water, the outlet water is sent for fine filtration, such as sand filtration, before being introduced to further treatment such as ion exchange. The filtration rate in the fine filtration unit of industrial filters is about 100 liters per square meter per minute. Pressure filters are widely used for rapid filtration in industry. In pressure-type filters, the available pressure usually ranges from 2 to 4 bar. Gravity filters are employed if a large amount of water is to be coagulated, settled and softened. The fine filtration unit consists of a layer of granular filter medium supported by a layer of graded gravel, and is equipped with the necessary pipework and fittings to carry out filtration and backwashing. Capacity of such units ranges from 10 to 1000 liters per minute (LPM). The system is normally designed for a standard flux rate of about 100 LPM per square meter of bed area. Height of the vertical filter is about 1500 mm. Horizontal filters are about 2500 mm in diameter, range from 3500 to 8000 mm in length and have capacities ranging from 600 to 1500 LPM. Single or multiple units are used to carry out filtration. The filtration is accomplished by passing the water through a bed of a fine granular filter medium supported by several layers of coarser material. Fine sand and anthracite are widely used filter media. The advantage of using a granular filter medium is that during backwashing, the bed expands and releases the insoluble matter collected during the filtration process. The filter continues to be operated until the gauges show sufficient loss of head to indicate that the filter unit should be backwashed. Backwashing is normally done for about 15 minutes when a head loss of 0.3 bar is observed. The minimum flow for backwashing is about 400 LPM per square meter of bed area.

The combination of sedimentation, coagulation, settling and filtration is widely used and gives very satisfactory results both for turbidity and color removal. Chlorination is widely used in filtration plants due to its sterilizing properties. It may be employed (i) before clarification, (ii) after clarification, or (iii) in certain cases both before and after clarification.

Activated carbon filters are used for removing taste and odor from water. The activated carbon filter is similar to the vertical pressure sand filter. It contains a bed of crushed and screened activated carbon ranging in depth from 600 to 1000 mm, supported by graded gravel and coarse sand. The filtration rate ranges from 150 to 250 LPM per square meter of bed area. Periodic backwashing is done to maintain the performance.

Iron is an objectionable constituent in water. For most industrial uses, it should not exceed 0.1 ppm. The removal of soluble iron from water is done by one of the following processes: (i) aeration, settling and filtration; (ii) sodium cation exchange (zeolite) water softening; (iii) hydrogen cation exchange; (iv) lime soda water softening; (v) two-stage lime and sodium cation exchange; and (vi) manganese zeolite process.

The removal of iron from water containing soluble iron (mostly in form of ferrous bicarbonate) is accomplished by oxidizing and precipitating the iron as ferric hydroxide, settling and filtration. Theoretically, 1 ppm of dissolved oxygen oxidizes 7 ppm of ferrous iron, expressed as Fe. Oxidation of iron is carried out in aeration equipment called an aerator or with chlorine as oxidizing agent.

Water containing iron in the form of ferrous bicarbonate can be removed simultaneously with the hardness by sodium cation exchange. The cation exchangers used are of high-capacity resin type, carbonaceous type or greensand type. The reaction involved in the removal of iron on the softening run is:

$$Fe(HCO_3)_2 + 2NaR = FeR_2 + 2NaHCO_3$$

where R is the complex radical of the cation-exchange resin.

The reaction involved in the regeneration with salt at the end of the run is:

$$FeR_2 + 2NaCl = 2NaR + FeCl_2$$

With the high-capacity resin type of cation exchanger, it is usual practice to limit the application of this process to waters containing not more than 0.5 ppm of iron, as Fe, for each 1 ppm of hardness present up to a maximum of 50 ppm of iron. In the removal of iron by sodium cation exchange, the raw water should not be allowed to come into contact with air before it passes through the water softening units, as ferric hydroxide will precipitate on and in the cation-exchange bed.

Manganous manganese, like ferrous iron, may be removed by sodium cation exchange simultaneously with the removal of the hardness, as per following reaction:

$$Mn(HCO_3)_2 + 2NaR = MnR_2 + 2NaHCO_3$$

The reaction involved in the regeneration with salt at the end of the run is:

$$MnR_2 + 2NaCl = 2NaR + MnCl_2$$

Ferrous iron, whether present as the bicarbonate, chloride or sulphate, can be removed by hydrogen cation exchange. As a rule, the removal of iron by this process is carried out simultaneously with the removal of calcium, magnesium and sodium. The reaction with ferrous bicarbonate takes place as follows:

$$Fe(HCO_3)_2 + 2HR = FeR_2 + 2H_2O + 2CO_2$$

The regeneration of the exhausted cation exchanger is carried out with sulfuric acid, which restores hydrogen to the exchanger and releases calcium, magnesium, sodium and ferrous iron as the sulphates. The reaction of the ferrous cation exchanger with the regenerant, sulfuric acid, is as follows:

$$FeR_2 + H_2SO_4 = 2HR + FeSO_4$$

Manganous manganese, whether present as bicarbonate, chloride or sulphate, can be removed by hydrogen cation exchange. Using the bicarbonate as an example, the reaction is as follows:

$$Mn(HCO_3)_2 + 2HR = MnR_2 + 2H_2O + 2CO_2$$

The regeneration of the exhausted cation exchanger is carried out with sulfuric acid, which restores hydrogen to the exchanger, removes the manganese as manganous sulphate and restores the cation exchanger to its original hydrogen state. The reaction of the manganese cation exchanger with the regenerant, sulfuric acid, is as follows:

$$MnR_2 + H_2SO_4 = 2HR + MnSO_4$$

Ferrous bicarbonate can be removed by any of the cold lime soda water-softening processes by oxidizing the iron by aeration so that it is precipitated as sludge as insoluble ferric hydroxide. The aeration is done in either an open or a forced draft aerator. As the pH of lime treated water is alkaline, around 10.0 or more due to the chemicals used to remove the hardness, the oxidation of iron is rapid and complete. Efficient removal of manganese takes place in any of the cold lime soda processes by simply aerating the water before it enters the softening equipment. Due to the high pH values during treatment, soluble manganese compounds are steadily oxidized to the insoluble manganic hydroxide.

Manganese zeolite is made from processed greensand zeolite by alternate treatments with manganese sulphate and potassium permanganate. This results in precipitation of the higher oxides of the manganese in and on the granules of greensand. In industry, manganese zeolite filters are used for removing up to 1 ppm of iron or manganese from water because potassium permanganate is a rather expensive material. The maximum flow rate in pressure units is limited to around 120 liters per square meter per minute of the bead area. In the manganese zeolite process, no softening of water is involved. The bed oxidizes the ferrous or manganous bicarbonates to the higher insoluble hydrated oxides and simultaneously filters them out of water. Backwashing and regeneration are carried out at specified intervals depending on the amount of iron or manganese present in water.

Ferric hydroxide was used as a part of one of the first processes for removal of silica from water. In the ferric hydroxide process, ferric sulphate is used in conjunction with lime at an optimum pH value of 9.0 to form ferric hydroxide. The dosage required depends on the silica content of the raw water and the extent to which it is to be reduced. In general, it is about 8 to 20 ppm of ferric sulphate for each ppm of silica removed. This process has been replaced by (i) basic anion exchangers, (ii) the cold lime soda magnesia water softening processes, and (iii) the hot lime soda magnesia water softening processes.

Some industries require total removal of minerals from water. The sodium cation-exchange (zeolite) process does not reduce the bicarbonate, sulphate or chloride content, which appears in the effluent as sodium salts. The hydrogen cation-exchange process removes the bicarbonate but does not affect the sulphate or chloride content, which appear as corresponding acid in the effluent. The cold lime soda process and the hot lime soda process reduce alkalinity, but have no effect in reducing the sulphate or chloride content or sodium alkalinity in water. The total removal of minerals from water is carried out by either ion-exchange demineralization or distillation. Both processes provide demineralized water. However, ion-exchange demineralization is quite popular in industry due to ambient temperature operation, particularly for surface water and low-salinity raw water. For high-salinity water, the cost of ion exchange increases with increasing salinity. For high-salinity water, distillation or other methods of desalination are quite promising.

Ion-exchange demineralization is carried out by removing the cations by hydrogen cation exchanger and anions by anion exchanger. Weakly basic anion exchangers and strongly basic anion exchangers are commonly used.

The weakly basic anion exchanger is used to remove strongly ionized acids but does not remove weakly ionized acids. At the end of each operating cycle, the weakly basic anion exchanger unit is backwashed, regenerated with a solution of soda ash (sodium carbonate) and rinsed back to service. As the weakly ionized acids are not removed by a weakly basic anion exchanger, the effluent contains the same amount of silica as the inlet water and carbon dioxide corresponding to that formed in its passage through the hydrogen cation exchanger plus its original carbon dioxide content. Carbon dioxide is removed by a decarbonator or deaerator.

The strongly basic anion exchanger removes both strongly ionized acids and weakly ionized acids. At the end of each operating cycle, the unit is backwashed, regenerated with a solution of caustic soda, rinsed and returned back to service. While the strongly basic anion exchanger can remove carbonic acid, it is cheaper to remove carbon dioxide content by mechanical rather than chemical means. Therefore, in most cases, the bulk of the carbon dioxide content of the acid effluent from the hydrogen cation exchanger used in the first step of the process is removed by a decarbonator or deaerator before entering to a highly basic anion exchanger unit.

Distillation is a well-known process for removing volatile liquids from non-volatile impurities. The quality of distilled water is excellent. It removes all the non-volatile impurities, both strongly ionized salts as well as weakly dissociated silicic acid. However, it is relatively expensive.

6.2.3 CASE STUDIES

6.2.3.1 Chemical Industries

Chemical industries are large users of water drawn from surface water and groundwater sources. If surface water contains turbidity, color or other impurities, it is treated for removal of the impurities to the level specified by the end-use requirement. Similarly, if groundwater contains impurities and contaminants, it needs to be pretreated followed by subsequent treatment as per the requirement.

Several plants use large quantities of steam for different purposes, such as heating, boiling, evaporation, distillation and other processing operations, as process steam or as a utility. Many industries generate their own steam at moderate pressure or high pressure depending on requirements. In plants operating their steam boilers in the moderate pressure range, sodium cation exchange is normally used to soften the water, preceded in some cases by the cold lime process to lower the alkalinity and total solids in the final effluent. Hydrogen cation exchange, with the acid effluent neutralized, is also used for the same purpose with the effluent from the sodium cation exchange. In case of steam-generation plants operating at higher pressures, hot water softening and silica removal processes, such as hot lime soda, two-stage hot lime and sodium cation exchanger, are normally used. For higher pressures, ion-exchange demineralization is used. The two-step hydrogen cation exchange and strongly basic anion exchange demineralization and silica removal system is widely used. Some of the chemical industries use recirculated high-temperature water. In such cases, sodium cation exchange is widely used for treating and softening the water.

Many chemical industries use cooling water as utility. If the cooling water is used once and discharged back, it may require only chlorination and acid treatment. If cooling water is recirculated through cooling towers, treatment to prevent scale and organic growth is done using the cold lime process, two-stage hot lime and sodium cation exchange, or the acid process. Chlorination of the recycled water may also be required. The LSI helps in formulating the degree of treatment.

Several industries use water as process water of a quality specified for their requirement, such as taking part in a reaction. The quality of water required may be as high as ultra-pure water. Process water in general must be clear, colorless, and free of inorganic and organic impurities. Whether or not further treatment is required depends largely on the quality required for its end use. For example, in pharmaceutical industries, the ion-exchange process is used to obtain the widely required distilled water.

In the refining of salt and manufacture of marine chemicals, the conventional treatment normally followed includes coagulation, settling, filtration, iron removal, lime and magnesia removal. The equipment employed may consist of a sedimentation basin, coagulation, settling and filtration equipment, cold lime process equipment, etc.

6.2.3.2 Textile Industries

Textile industries use lots of water during manufacturing. The water used in the wet processing of textiles should be free from color, iron, manganese, turbidity, hardness and organic growths, and should not be corrosive. Clarification normally used to deal with turbidity and color issues involves sedimentation, coagulation, settling and filtration through gravity or pressure sand filters. Aluminum sulphate is normally used as coagulant. For the removal of color, clay is frequently employed as an aid to floc formation. The presence of iron and manganese in water leads to staining and catalytic effects. The removal of iron, when it is present as ferrous bicarbonate, is accomplished by aeration, preferably at alkaline pH (pH > 7), settling and filtration. Manganese, when present as manganese bicarbonate, is removed by aeration, settling and filtration. Aeration is carried out at pH values above 9.0. Organic iron and manganese normally respond to aeration, coagulation with alum, settling and filtration. Hardness in water is not desirable as it causes several kinds of problems in wet processing operations. In silk processing, soaking in hard water increases reel breakage. In wool processing, rinsing and dyeing operations are best conducted in zero-hardness water. Sodium ion exchange and ion-exchange demineralization are widely used in the treatment of process water for manufacture of synthetic fibers. In most textile plants, the boiler feed water is softened by sodium cation exchange.

Algae, crenothrix and slimes are objectionable as they may cause clogging, staining, odors and wastage. Chlorination is widely used for their prevention. Copper sulphate treatment is carried out to inhibit algae and other growths. Corrosion in the process water level is usually inhibited by caustic sodium silicate treatment.

6.2.3.3 Food Industries

Different qualities of water are used for different purposes such as cooling, washing, flushing, processing, boiler feed and other general uses. The requirements for boiler feed and cooling waters are the same as in other industries. Water used for cooling, flushing and other general purposes should be clear, colorless, free from iron and manganese, and free from objectionable tastes and odors. Water for washing operations should preferably be of zero hardness. For process water, the general specifications are clear, colorless, free from iron and manganese, and free from objectionable tastes and odors, even when food products are put through a sterilizing operation. However, the specific quality requirements for process water depend on the end use and type of food industry.

The water used in bakeries should be clear, colorless, odorless and free from iron and manganese. The presence of calcium salts is necessary for proper fermentation. For cake baking, soft water is preferred as it yields a better and more uniform product. Zero-hardness water is preferred for cleaning purposes in bakeries and sodium cation exchanger water softeners are widely used. Hydrogen cation exchange and hot lime soda process are also used. Bakeries also use coagulant feeds, pressure sand filters, and iron and manganese removal equipment.

The water employed in dairies should be free from iron, manganese, taste, color, odor and hardness. It should be bacteria free. Hard water has potential to cause scaling in boilers, pumps, pipes and fittings. It is troublesome in pasteurizers, as the temperature of the water used is high. Coagulation, settling, filtration and chlorination followed by water softening is carried out for the removal of suspended solids, turbidity, color, odor, taste, iron, manganese and hardness. A zeolite water softener is usually employed in dairies.

Ice is manufactured in several food industries and cold storage plants. The water used in ice making should be absolutely clear, odorless, colorless, tasteless and free from iron and manganese. Bicarbonate hardness is objectionable, as both calcium and magnesium are precipitated on freezing. The water treatment process selected mostly depends on the quality and composition of the raw water. If total salt content in raw water is low, the sodium cation-exchange process is employed for treatment. The cold lime process is used to reduce bicarbonate hardness. For water high in sodium bicarbonate, hydrogen cation exchange is used, as it removes the bicarbonate. Ion-exchange demineralization is also used to obtain excellent quality ice. Where surface water is used as the raw water, coagulation, settling and filtration is carried out followed by chlorination as required.

6.2.3.4 Electroplating Plants

Water in electroplating plants is used for (i) cleaning, (ii) rinsing after cleaning, (iii) making up the plating bath, and (iv) rinsing after plating. While organic solvents are used for cleaning the metal before plating, the cleaners commonly used are alkaline in nature, such as resin soap, soda ash, caustic soda, etc. As hard water produces precipitates with alkali, water of zero hardness is recommended. Otherwise, precipitate may adhere to the metal and cause pitting in the plating.

The water used for rinsing after the alkaline cleaning should also be of zero hardness. Otherwise, it will react with the strongly alkaline water and form precipitate. In acid plating, where sulphates and sulfuric acid are used, concentration in the plating bath may lead to the formation of calcium sulphate scale. In the final rinsing of the metal after plating, hard water frequently causes streaks.

Carbon dioxide in water may cause embrittling and edge cracking in nickel plating. Turbidity in water is objectionable. Chlorides, sulphates and carbonates are harmful in silver plating. Hence demineralized water is preferred. In chromium plating, water demineralized by ion exchange is widely used for the plating bath and the rinse water.

Water treatment methodology used in electroplating industries generally involves coagulation, settling, filtration, aeration, chlorination, iron and manganese removal, softening and demineralization.

6.2.3.5   Pulp and Paper Mills

Pulp and paper mills use a huge amount of water. The quantity depends on the type of mill and product manufactured. Free carbon dioxide adversely affects sheet formation in paper machines. Hydrogen sulfide is corrosive to metals and even a small amount present in water may reduce the life of the wire on cylinder machines. It is advisable to remove it by chlorination even after deaeration. Microorganisms are not desirable. Odor should be avoided, particularly when making paper products for wrapping or food packaging. The choice of water treatment depends on the type and extent of impurities present in raw water.

6.2.3.6   Iron and Steel Mills

Large quantities of water are used in iron and steel mills for different purposes, such as cooling operations, pickling, rinsing, plating, boiler feed and general uses. Cooling water is generally carried out by sedimentation, chlorination, coagulation, settling, filtration, iron and manganese removal, and softening. Boiler feed water treatment may involve sodium cation-exchange water softener, two-stage cold lime and sodium cation-exchange water softeners, and hot lime soda water softeners.

6.2.3.7   Distilleries

Water consumption in distilleries varies depending on the types of stills. The quality of water to be used in distilleries need to be colorless and free from undesirable constituents such as odor, iron and manganese, and should be of approved bacteriological quality. A small amount of bicarbonate hardness is usually desirable in the water for maintaining the proper pH value during fermentation. Care is taken to avoid scale formation due to water quality in the condenser of the beer still and column still. However, some distilleries do not treat the water used in mash coolers even when it is quite hard, but rely on periodic cleaning of scale from the coils.

6.3   WATER TREATMENT FOR RIVER REJUVENATION

In many parts of the world, rivers are polluted and river water is not of potable quality. At times, water in the river is not even of bathing quality.

The Ganga river basin is the biggest in India and accounts for more than 25% of the country's total geographical area. The river traverses a course of more than 2500 km through the plains of northern and eastern India, originating in the Himalayas and flowing into the Bay of Bengal.

Increasing urbanization, industrial growth and population have significantly impacted the water quality of the river Ganga, particularly during the dry season. The primary sources of pollution are untreated sewage and industrial wastewater. Each day, more than 500 million liters of wastewater from industrial sources are dumped directly into the river. In some places, the wastewater entering the river is raw and untreated. This industrial waste is making the Ganga dangerous for drinking or bathing due to the presence of chemical contaminants and heavy matter. For example, a tannery industry situated along the river produces millions of liters of wastewater mostly consisting of effluent from tanneries. These pollutants entering the river are disturbing the natural ecosystems. High levels of toxic chemicals such as chromium, arsenic, cadmium and lead are present in the polluted river water, adversely affecting the health of those using this water for drinking, cooking and bathing. It is estimated that 764 industries along the Ganga main stream consume more than 1100 million liters per day (MLD) of water and discharge 500 MLD of wastewater (Table 6.8). Sugar, pulp, paper, textile and distillery industries are responsible for about 70% of the pollution, whereas tanneries are responsible for the most harmful types of toxins.

TABLE 6.8

Tentative Water Consumption and Wastewater Generation in Industries along the Ganga River

S. No.	Type of Industry	Total Units	Water Consumption (MLD)	Wastewater Generation (MLD)
1	Tannery	444	28.7	22.1
2	Pulp and paper	67	306.3	201.4
3	Sugar	67	304.8	96.0
4	Textiles, bleaching and dyeing	63	14.1	11.4
5	Chemicals	27	210.9	97.8
6	Distillery	33	78.8	37.0
7	Food, dairy and beverage	22	11.2	6.5
8	Others	41	168.3	28.6
9	TOTAL	764	1123	501

Pollution from industries constitutes around 20% of total pollution load by volume. However, its contribution to polluting the river is much greater, due to the higher concentration of pollutants. Two types of methodologies can be used for treatment:

1. Treatment of industrial wastewater in the common effluent treatment plant (CETP).
2. Decentralized treatment of industrial wastewater at individual plant level based on effluent characteristics.

Membrane-based effluent treatment system either as a standalone process and/or integrated with other primary and secondary processes can play a vital role in controlling critical effluent and discharge parameters in accordance with environmental norms. The effluent treatment system is designed, based on the effluent characteristics, to achieve safe discharge. Ultra-filtration membrane-based water clean-up plants for turbidity, pathogenic and organic removal appear promising. These technologies are scalable at both higher and lower capacities.

Water discharge from industries needs to be analyzed in detail to identify the contaminants so that membranes/technologies can be screened for that particular industry. Once the specific effluent characteristics from that industry are identified, an appropriate system based on conventional and membrane technology is designed. In any membrane process, the treatment of effluents will generate two streams: one which is fit to be discharged as such and another concentrate/reject stream (of low volume) which will contain all the contaminants in higher concentrations than the feed effluent streams. The concentrate/reject from membrane processes has no process-added chemicals and generally reflects the characteristics of the feed effluent as it simply redistributes the contaminants. Management of the concentrate/reject from membrane processes depends upon the feed effluent water composition and the membrane process being used. This reject/concentrate slurry emanating from individual types of industry like tannery, textiles, pulp and paper, etc., will have to be treated in different ways. The membrane effluent/reject treatment, though low in volume, is an involved and potentially cost-intensive process that cannot be generalized, as it is industry specific. The authorities involved may use standard processes like chemical post-treatment, shallow solar evaporation ponds, incineration, deep-well injection, etc.

REFERENCES

1. Khuntia, S., 2006. *Desalination and Water Purification Technologies*, Bhabha Atomic Research Centre, Mumbai, India. Water purification technologies – A position paper.
2. Nordell, E., 1961. *Water Treatment for Industrial and Other Uses*, Reinold Publishing Corporation, New York.
3. World Health Organisation, 1993. *Guidelines for Drinking-Water Quality*, 2nd Edition, Vol. 1 - Recommendations, WHO, Geneva, pp. 172–181.

7 Wastewater Treatment, Recycling and Reuse

7.1 WASTEWATER TREATMENT

Processes and systems for wastewater treatment can be classified in a number of ways. It is quite common to characterize them by function such as precipitation, bioremediation, adsorption, etc. Primary, secondary and tertiary treatment methods used include sedimentation, biological treatment with sedimentation, and removal of residual or nonbiodegradable constituents. During primary treatment, grit and suspended matter is removed. After the physical separation process, the wastewater is biologically treated in the trickling filters to reduce oxygen demand, which can be biological (BOD) or chemical (COD). Clarification is carried out in the clarifier, after which the secondary-treated wastewater goes for tertiary treatment, where residual and nonbiodegradable constituents are removed. Chlorination, ozonation or organic biocide addition are performed to disinfect the water. In view of the hazardous effects of chlorine products, there is a need to develop oxidants and biocides that are environmentally friendly and pose no health hazard.

The majority of water contaminants, including colloids, particles, bacteria and pyrogens (bacterial fragments), are negatively charged. Zeta potential-based filtration is a promising technique for removing water contaminants. Properties of the filter fiber, including shape and surface area, affect filtration efficiency, but surface charge has the most significant impact. Surface charge is typically characterized by zeta potential. Filter media can be chemically modified to have a positive zeta potential. The removal mechanism is electrostatic attraction, which is effective in water over the typical deionized water pH range of 5–8. Some of the positive zeta-potential elements additionally remove fine negatively charged organisms and particles well below their micron range. Changing the zeta potential between the particles and the fibers by adjusting the pH or ionic concentration can have a significant effect on filter efficiency. The efficiency of removal by electrostatic attraction reduces as the active sites of the filter fill up with collected particles. However, actual efficiency does not drop below the removal rating of the filter. Because of the highly porous nature of the membrane, the actual total membrane area containing positive zeta-potential sites is orders of magnitude greater than the effective filtration area. Thus, the capacity for electrostatic adsorption of fine particles is quite large.

Nylon 66, cellulose fibers and different varieties of organic polymers are being used as zeta-potential high area filter elements for microbial and particulate removal. Though many zeta potential-based filters are being used for water treatment, there is scope for development of more efficient filters. Nanomaterials have novel and unique physical and chemical properties due to quantum confinement and large surface-to-volume ratio. Ceramic oxides, metals and metal oxides, sulfides, nitrides, carbon including fullerenes, layered chalcogenides such as TiO_2, Fe_2O_3, Si, Ag, CuS, Ag_2S and AgBr, etc. are a few examples of nanostructured materials that can be explored for their suitability as zeta-potential filters.

Zeolite-based inorganic ion exchangers and filters have high potential for water treatment because zeolite is an important class of compounds known to act as inorganic ion exchangers/molecular sieves. They are environment friendly, available and quite economical. However, their ion-exchange capacity is relatively poor compared to synthetic organic ion exchangers. This can be overcome if the particle size of these materials is brought up to nano-size. As the surface area-to-volume ratio increases, the capacity increases correspondingly.

Ion-exchange resins with phosphorus, arsenic and organic molecules imprinted in the resin matrix have potential to compete with the existing sulphonic/quaternary ammonium hydroxide ion-exchange resins. They are specific to certain metal ions.

Eutrophication is an enrichment of plant nutrients such as nitrogen and phosphorus in water bodies. The process takes place through natural or manmade activities such as the addition of industrial waste, domestic effluent, etc. It promotes the production of phytoplankton, algal blooms and aquatic vegetation, including water hyacinth, aquatic weeds, water fern and water lettuce, which in turn provide ample food for herbivorous zoo-plankton and fish. As a result, the aquatic oxygen level decreases, aquatic organisms begin to die and the clean water turns into a stinking drain. The problem can be addressed by removing plant nutrients and the products of eutrophication effects. Chemicals like $CuSO_4$ have been used to solve the problem but these solutions are only temporary and may have adverse effects on fish. A large proportion of eutrophication-generated waste is organic in nature and can be decomposed by saprophytic bacteria (which are harmless and feed on dead organic matter) or microorganisms. Denitrification microorganisms can convert nitrates into nitrogen gas which is released from water bodies. Orthophosphate is assimilated into cellular adenosine triphosphate (ATP) by the competitive population of microorganisms. Microorganisms similar to saprophytic bacteria can be used as a bioremediation tool to address eutrophication problems. In another non-chemical water treatment method derived from soil biotechnology, organics and oxidizable inorganics are processed using fundamental natural chemical reactions such as respiration, mineral weathering and photosynthesis. This green engineering approach to wastewater treatment and recycling is based on bio-conversion and ensuring compatibility of effluent with aquatic life.

Wastewater treatment requirements for common effluents from industrial manufacturing processes are normally laid down by federal and state agencies. Treatment of wastewater involves a number of steps. Quite often, wastewater streams that pose problems are relatively small. It makes sense from the overall treatment and techno-economics point of view to separate the problem stream from the overall flow.

Table 7.1 lists treatment methods for the removal of undesirable constituents from wastewater.

In conventional wastewater treatment, the effluent is normally discharged to the environment in accordance with effluent discharge standards, and the appropriate treatment method is used to deal with undesirable constituents. With increased awareness of environmental considerations and increasing demand for water, treating the wastewater and improving its quality for reuse and recycling is becoming

TABLE 7.1

Wastewater Treatment Processes

S. No.	Purpose	Treatment Method
1	Suspended matter and turbidity removal	Storage, sedimentation, coagulation, filtration
2	Colloidal matter removal	Coagulation and filtration
3	Color removal	Coagulation, sedimentation, filtration, chlorine, ozone, activated carbon
4	Odor removal	Aeration, activated carbon adsorption
5	Taste removal	Activated carbon adsorption
6	Free carbon dioxide removal	Aeration
7	Iron and manganese removal	Aeration, filtration, ion-exchange process
8	Hardness removal	Lime treatment, ion-exchange process
9	Oil and grease removal	Filtration, activated carbon adsorption
10	Biological contaminants disinfection	Chlorination, ozonation, copper sulfate treatment, bromination, ultraviolet rays
11	Dissolved salts removal	Ion-exchange process, membrane process
12	Dissolved organic removal	Chemical oxidation, activated carbon adsorption, membrane process

increasingly important. This approach is reinforced by the need to follow a zero-liquid-discharge (ZLD) policy wherever feasible.

Membrane processes have high potential for deployment in the treatment of wastewater and industrial effluents. They help remove pollutants and contaminants from the effluent stream, making it safe to discharge into the environment, and support recovery of a significant proportion of good-quality water for recycling. Water recovery and reuse with the potential for zero effluent discharge is gaining ground. Recycling and reuse appear promising due to scarce availability of good-quality water and rising costs. Plans to treat wastewater from office and residential buildings for recycling and reuse are under consideration worldwide. An increasing number of water reclamation systems have been coming into operation every year. Wastewater from the wash basins of a building, which contains little in the way of organic contaminants, is classified as low BOD wastewater, whereas wastewater from the kitchen is high in organic contaminants, containing BOD and oily substances, and is therefore classified as high BOD wastewater. Sewage wastewater is also high in organic contaminants, containing ammoniacal nitrogen and coloring matter. Major treatments adopted in individual buildings are biochemical methods, including activated sludge and contact oxidation, and physical methods, such as ultrafiltration.[1]

Microfiltration (MF), ultrafiltration (UF), nanofiltration (NF) and reverse osmosis (RO) are common pressure-driven membrane processes that are quite useful for treatment of wastewater. The suspended solids in wastewater are removed by MF. Removal of suspended solids and large impurities prior to fine filtration lengthens the life and decreases the fouling of the associated membrane system. Ultrafiltration (UF) is useful for separating macromolecules and submicron particles, including oil emulsions and very large molecules such as polymeric compounds having molecular weight of 1000 and above. Nanofiltration is capable of separating molecules ranging from 300 to 1000 molecular weight. An NF membrane shows excellent solvent stability and a wide operation pH range (0–14). It also helps with selective separation of low molecular weight organics from saline solution.

An RO membrane has a very small pore size in the range of 5–10 A°. Ions and molecules having 100 and above molecular weight can be easily removed by RO. No two wastewaters are exactly alike in quality. The quality of the wastewater differs depending on the type and source of the effluent. It is necessary to carry out laboratory and pilot tests to determine flux rates and performance under different conditions of temperature and pressure. The feed under pressure is passed through the rig and permeate flow is noted for each feed against pressure and time. It is desirable to study the fouling potential under operating conditions as well as establishing the best membrane cleaning procedure. Product quality and permeate flux are evaluated. The membrane system giving the desired performance and quality is selected.

7.2   RECYCLING AND REUSE

Water recycling generally refers to the internal use of wastewater by the user before it is discharged to a treatment system or other point of disposal. The term 'water reuse' normally applies to wastewater discharged from municipalities and industries, and then drawn by other users. After treatment, reclaimed water is used for industrial applications, cooling, irrigation, etc. Water recycling and reuse play an important role due to the scarcity of water availability in many places, which in turn is due to improved quality of life, population increase and industrial growth. In several areas, there is a need to achieve 100% wastewater treatment, with more stringent standards, so that water can be recycled and reused. It guarantees the availability of water for process needs as well as other uses and brings down the raw water requirement. It is also possible to make additional savings through recovery of valuable by-products from the concentrate stream. Water recycling and reuse help ensure compliance with pollution control and clean environment regulations by reducing effluent discharge. In some cases, water recycling also considerably reduces capital investment in water treatment for new projects. Water recycling and reuse is an area where tremendous scope as well as opportunity exists.

Screening, settling, clarification, chemical treatment, chlorination, ultraviolet disinfection, hydrogen peroxide disinfection and ion exchange are some of the well-known wastewater treatment processes. Advanced aerobic and anaerobic digestion technologies have made effluent recycling in a single process a reality. Microfiltration, ultrafiltration, nanofiltration, RO, membrane bioreactors and photochemical oxidation have the potential to play an important role in water recycling and reuse.

7.2.1   PHOTOCATALYTIC METHOD

The photocatalytic method is quite promising for the treatment of water containing large amounts of organic impurities. Sunlight in the presence of a zinc oxide or titanium dioxide (TiO_2) catalyst is effective in converting organic impurities into carbon dioxide. However, the efficiency of the treatment process is limited because only the ultraviolet (UV) component of the electromagnetic spectrum of solar light helps in the conversion. Only 5% of sunlight is emitted as UV light. Thus, the visible component of the sun's energy (95%) does not contribute to this reaction. Among semiconductors, TiO_2 is regarded as a promising photocatalyst for the UV-assisted photocatalytic process. The TiO_2 band gap is 3.2, eV which makes it photoactive only in ultraviolet regions. The doping of this oxide with other transition metal ions (Fe, Zn, Co, etc.) leads to a red shift in the band gap transition. This can make the photocatalysis reaction take place even with the visible component of solar energy which is available in abundance. Further, the effectiveness of the process can be increased with the use of nano-catalysts or Q-sized particles (TiO_2). Use of Q-sized particles increases the efficiency of the process of decomposition through the large increase in surface area of the catalyst and the increase in the recombination time of electrons and holes. In order to make the process feasible for large-scale treatment of industrial effluent, the catalyst has to be supported. The selection of support material depends, among other things, on the design of the plant. Several candidate support materials, such as zeolites, can be tested for this type of application. It requires:

a. Synthesis of Q-sized semiconductor materials (like TiO_2) with improved methods to obtain them at reasonable cost and good quality.
b. Study of the effect of transition metal ion and other metal ion doping in Q-sized semiconductors on the catalytic efficiency of treatment under a solar light source.
c. Methods of developing support materials for these catalysts, and studying their efficiency in the catalytic process.

7.2.2 BIOTECHNOLOGY

Significant progress has been made over the years in water purification systems derived from biology and engineering principles.[2] Biotechnological methods of water purification fall into the following categories:

(i) microbial techniques for wastewater reclamation,
(ii) phyto-techniques for wastewater reclamation,
(iii) bioremediation of groundwater.

In the microbial techniques, microbes are divided into three groups: bacteria, fungi and actinomycetes. Bacteria are believed to be major agents for breaking down organic pollutants and for absorbing inorganic pollutants in wastewater. Corynbacterium in an aquatic environment seems to be readily adaptable to heterocyclic compounds and hydrocarbons. Fungi appear to have a greater ability to degrade hydrocarbons in wastewater. Actinomycetes are known to attack a wide variety of organic pollutants including phenols, pyridines, glycerides, paraffins, steroids, chlorinated and non-chlorinated aromatic compounds.[3] White rot fungi are common forest organisms causing characteristic decay of many tree species. One of the exciting prospects from

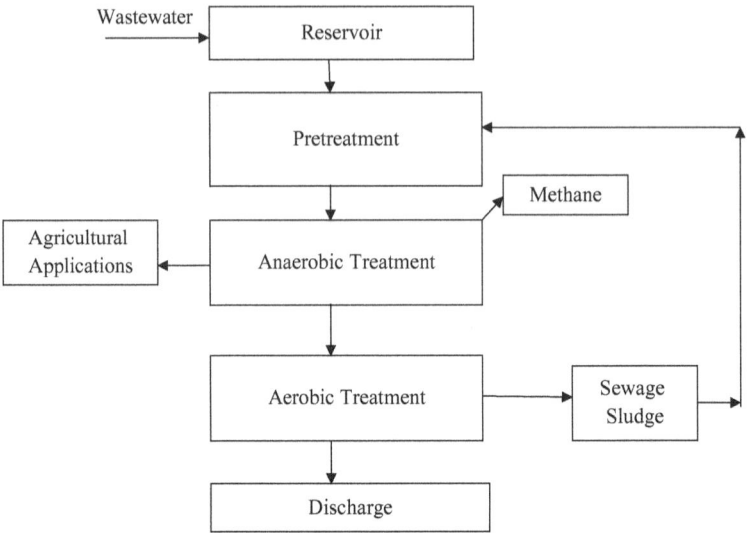

FIGURE 7.1   Anaerobic/aerobic biotreatment method.

biotechnology has been the use of anaerobic organisms that have a broader range of biodegradation capabilities. In order to enhance the efficiency of wastewater treatment, aerobic and anaerobic microorganisms are sequentially added to a treatment system. A flowchart of anaerobic/aerobic biotreatment methods is shown in Figure 7.1.

Some plant roots and root exudates degrade or absorb pollutants in wastewater, purifying the water.[4] Studies have shown that coniferous and hardwood forest ecosystems can be effective filters for improving the quality of wastewater by removing nutrients, metals and organic pollutants. Crops like soyabean (*Glycine max Nerr.*), sunflower (*Helianthus annuus L.*) and rye (*Secale cereale L.*) can remove pollutants such as nitrogen, phosphorus, selenium, molybdenum and nickel.[5]

Heavy metals such as mercury, zinc, lead, arsenic, cadmium, copper etc. in wastewater can be absorbed and organic pollutants in wastewater can be degraded with the growth of microalgae. Proliferation of microalgae in wastewater promotes photosynthesis. Oxygen from photosynthesis can be provided for microorganisms oxygenating organic pollutants in wastewater. Carbon dioxide from oxygenation can be used in the photosynthesis of microalgae. In these processes, the pH of the wastewater rises to lethal level for viruses and pathogenic bacteria.[3] The culture of microalgae in wastewater is an important approach for purifying wastewater.

The biodegradation of organic pollutants in wastewater is enhanced in the presence of plant roots and plant root exudates. Diverse species of heterotrophic microorganisms in the rhizosphere are brought together at high population densities, which may enhance stepwise transformation of organic pollutants by consortia or provide an environment conducive to genetic exchange and gene rearrangement.

Some forest species have the ability to purify wastewater, but uptake is low. Selection of forest species and the concentration control of pollutants in wastewater is a major design constraint. Artificial wetland biotreatment is another technique, in which an artificial wetland is gradually formed with the discharge of wastewater into grassland. For spray irrigation systems, a plant species or mixture can be chosen from water-tolerant perennial grasses.

Bioremediation methods for groundwater contaminated with organic compounds, metals and radionuclides fall into three categories: (i) biodegradation of organic compounds, (ii) biosorption and bioaugmentation of metals from solution, and (iii) in-situ mobilization and immobilization of metals in subsurface environments, where the microbial population can be designed to alter pH or redox potential through their metabolism. Rapid developments in biotechnology offer considerable potential for improving the performance of wastewater pretreatment systems. Microbial techniques and phyto-techniques are promising and need further investigation. There is scope to carry out further studies in the following areas:

1. Pretreatment techniques for conditioning water before using microorganisms on water bodies.
2. The search for effective and selective microorganisms.
3. Methods for continuous monitoring of dissolved oxygen, BOD and plant nutrients.
4. The search for economically viable and eco-friendly aquatic plants that can grow after bioremediation to prevent further eutrophication-generated ill effects.

The reuse of biologically treated wastewater is an effective strategy for water conservation, pollution control and recovery of valuable materials present in wastewater. At times tertiary treatment is incorporated.

7.2.3 MEMBRANE PROCESSES FOR WATER RECYCLING AND REUSE

Industrial wastewater has variable physical and chemical features due to the presence of different groups of organic and inorganic compounds, such as:

- soluble organics
- synthetic chemicals
- inorganic toxic acids
- heavy metals
- alkalinity and acidity
- oils, fats, colloids and suspended solids
- nutrients
- color, turbidity and odor

These pollutants are mainly removed by primary and secondary treatments. However, a tertiary treatment is needed to:

- comply with restricted conditions of a biological process, such as heavy metals less than 1 ppm with pH value in the range of 6–9
- remove non-degradable chemicals or species
- lower gaseous emissions

TABLE 7.2

Membrane Processes for Wastewater Treatment

S. No.	Membrane Process	Pollutant Removal Capability
1	Microfiltration	Turbidity
2	Ultrafiltration	Bacteria, virus, macromolecules
3	Nanofiltration	Bivalent ions and beyond, hardness, pesticides
4	Reverse osmosis	Monovalent ions, nutrients

Membrane processes are well established in the field of desalination, dialysis, etc. Their extension into environmental engineering and wastewater treatment is logical. Table 7.2 lists some of the membrane processes useful for wastewater treatment.[6]

Desalination technologies are largely used for producing clean water for human consumption and high-end industrial applications. In some cases, purified water is also used for agricultural and recreational activities. In such cases, membrane processes can be used to treat wastewater generated from process water and human consumption. The treated water can be used for agricultural and recreational purposes. Thus, desalinated water is used for high-end industrial and community water requirements and then reused for less restricted municipal and agricultural applications. It is then discharged to the environment in the normal way. Figure 7.2 illustrates a water reuse management scheme. This approach can lead to efficient resource utilization and better economics.

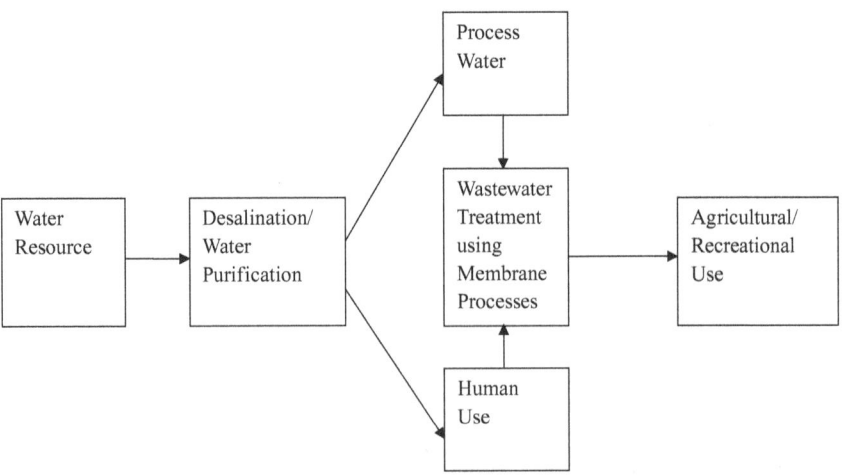

FIGURE 7.2   Water reuse scheme.

7.3 INDUSTRIAL WASTEWATER

Treated wastewater effluent can be more promising than traditional water resources in water-stressed areas. The suitability of wastewater for reuse depends on several factors:

- Presence of suspended solids
- Dissolved substances
- Biodegradable species
- Presence of heavy metals and toxic species
- Microbiological contamination

As a consequence, tertiary treatment is needed. In countries with arid and semi-arid climates, a series of stabilization ponds are provided. This solution has some drawbacks, such as algae growth, high suspended solids content and significant oxygen consumption. Membrane processes appear to be quite promising for tertiary treatment and polishing for reuse.[7] The choice of type of membrane technology depends on the quality and quantity of water to be processed. A general plan for membrane-based wastewater tertiary treatment is shown in Figure 7.3.

The type of membrane separation technology to be used depends on the quality of the wastewater as well as the quality of purified water required for reuse. Commonly used membrane separation technologies are microfiltration, ultrafiltration, nanofiltration and RO, which may be used as standalone technologies or in combination. For example, a combination of microfiltration and nanofiltration generally requires pretreatment of the nanofiltration feed to lower the silt density index (SDI). An RO membrane is recommended if wastewater feed contains dissolved solids to be removed. Microfiltration removes suspended solids and turbidity. Ultrafiltration is capable of removing bacteria and viruses. Nanofiltration has the capability to remove organics and multivalent ions.

Water recovery and recycling has the twin advantages of producing fresh water at a nominal cost while also addressing the problems of effluent disposal. The membrane process can play an important role in water recycling. As the membrane-based plant is modular in nature, industrial effluent can be treated individually or collectively, so

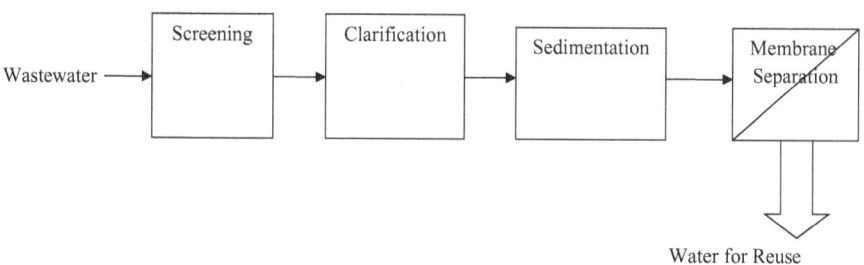

FIGURE 7.3 General plan of membrane-based wastewater tertiary treatment.

that the concentrate can be either reused in the process or other valuable components besides fresh water can be recovered from it. This concept needs to be explored on a case-by-case basis. ZLD can be achieved using a combination of membrane and thermal technologies.

7.4 CASE STUDIES

7.4.1 WASTEWATER IN THE TANNERY INDUSTRY

7.4.1.1 Recycling and Reuse in the Tannery Industry

Leather production requires a large amount of water: up to 100 L per kg of hides. Wastewater is generated in various operations, such as desalting, pickling, tanning, etc., that take place in tanneries. Surface water around the tanneries may suffer from high pollution levels. Efforts to improve treated water quality include tertiary and quaternary physico-chemical treatments. A microfiltration and a nanofiltration unit can be integrated with a biological treatment process for wastewater. The SDI of treated water after microfiltration will be about 4–5, which is the same as that of natural water normally used in tanneries. More than 70% of the overall water of the biological plant is recovered at volume concentration ratios (VCR: ratio between feed and retentate volume) of 9–10 in microfiltration step at 1.2–1.3 bar pressure with an average permeate flow of 40–50 liters per square meter per hour (LMH). Nanofiltration with VCR of 2–6 operates at about 10 bar pressure with permeate flux of 20–35 LMH. Table 7.3 shows the tentative quality of water produced by microfiltration and nanofiltration of biologically treated tannery wastewater. The permeate COD value depends on the VCR. The higher the VCR, the higher the permeate COD. The permeate flux decreases with the duration of operations and recovers substantially after daily membrane washing and cleaning with acidic solutions and surfactants. The quality of the water produced after nanofiltration is quite high and suitable

TABLE 7.3
Tentative Quality of Water Produced by Microfiltration and Nanofiltration of Biologically Treated Tannery Wastewater

S. No.	Parameters	Biologically Treated Wastewater after Settling as Feed	Microfiltration Permeate	Nanofiltration Permeate
1	pH	7.65	7.65	7.65
2	Chemical oxygen demand (COD) as ppm	260–330	220–240	50–120
3	Suspended solids (SS) as ppm	10–40	Not detectable	Not detectable
4	Chlorides as ppm	2100–2300	2100–2300	1800–2100
5	Sulfate as ppm	1500	1500	75
6	Ammonia as ppm	55	20	12

for recycling as well as agricultural use. It is free from microbial contamination and contains low levels of multivalent heavy metal contamination.

7.4.1.2 Recovery of Valuable Materials in the Tannery Industry

Using a membrane process for industrial wastewater treatment generates two types of streams: filtrate (permeate) and concentrate (retentate). The filtrate or permeate stream is good-quality water which can be reused in the plant. The retentate or concentrate stream contains valuable materials such as chemicals and by-products. It is possible to recover valuable materials from the concentrate stream through the membrane process.

In tanneries, animal hides are treated with inorganic and organic chemicals at different stages from storage and conservation through to dyeing and finishing. A lot of water is used and at the end of the process, water carries some chemicals as contaminants and pollutants. One of the chemicals widely used in the tannery industry is chromium sulfate, because it has a high penetration rate into the fibrillar interspaces of the skin and gives hides good mechanical and hydrothermal resistance and exceptional suitability for dyeing. As it is used to excess, about 30–40% of initial chromium is found in the wastewater and sludge of the biological waste treatment plant and tanned skin residues. Recovery of chromium from the tanning bath makes economic sense. Once the chromium has been recovered, the biological sludge and hide residues can easily be reused in agriculture. A process flow diagram for recovery of chromium solution from the tanning process is shown in Figure 7.4.

Table 7.4 gives some of the typical chemical composition of wastewater, ultrafiltration permeate and nanofiltration permeate.[8] Ultrafiltration operates at about 1–3 bar pressure with permeate flux of about 25 LMH. It gives good rejection of oil and fats. Nanofiltration operates at about 14 bar with initial permeate flux of about 25 LMH, reducing with time. Appropriate membrane cleaning is required at regular intervals to maintain the membrane flux. During ultrafiltration, more than 40% of the chromium in feed is recovered and 60–70% of the process water is recovered. Permeate is recycled back as process water.

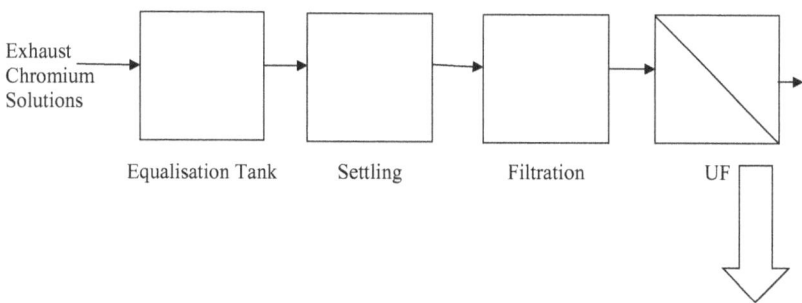

Chromium Sulphate Solution (concentrated)
for reuse in retanning and make-up tanning bath

FIGURE 7.4 Process flow diagram for recovery of chromium solutions from tanning process.

TABLE 7.4

Typical Parameters of Tannery Wastewater, Ultrafiltration Permeate and Nanofiltration Permeate

S. No.	Parameters	Wastewater Feed	Permeate from Ultrafiltration	Nanofiltration Permeate
1	pH	3.7–4.2	4	4
2	Oil and fats	300	2	Not detectable
3	Total suspended solids (SS) as ppm	700–2900	25	21
4	Sulfate as ppm	22000–23000	22000–23000	700
5	Chromium (III) as ppm	3600	2300	Not detectable

The overall scheme results in:

- recovery of chromium salts, chlorides and sulfate
- recovery of process water
- reduction of salinity (chloride and sulfate) in the biological wastewater treatment plant
- reduction of the chromium content in biological sludge and hence possibility of reuse of sludge in agriculture

7.4.2 RECYCLING AND REUSE IN THE TEXTILE INDUSTRY

The textile industry is known for its intensive water requirement. Conventional wastewater treatment schemes fully mix the various wastewater streams from different operations. The composition of the resulting wastewater stream is very complex, containing dyes, organics and salts. This also provides an opportunity to recover many of these species and reuse them. Membrane processes integrated with biological treatment have the potential to recover up to 90% of the water and 95% of the chemicals.[9] Recovery and reuse of water, colors, sizing agents and sodium hydroxide have been thoroughly assessed. Recovery of salts such as sodium chloride, sodium sulfate, lanoline and sericin (a silk protein) is under evaluation. Table 7.5 gives some of the typical parameters of wastewater or textile effluent.

Figure 7.5 shows a scheme for the treatment and recovery of textile effluent. Recovery and recycling of good-quality water and upgrading the biological water treatment of plant effluents to fit quality criteria for water reuse in agriculture, community and other applications offer a viable treatment, recycling and reuse alternative.

Recovery and recycling of good-quality water and recovery of valuable materials result in:

- saving of raw water due to recycling and reuse of water
- saving on biological treatment cost due to effluent volume and COD reduction
- recovery of valuable materials

TABLE 7.5
Typical Parameters of Textile Effluent

S. No.	Parameters	Overall Wastewater	Output from Biological Treatment Plant	Water from Membrane Processes
1	pH	2–13	7–8	7–8
2	Temperature (°C)	30–80	25	25
3	Chemical oxygen demand (COD) as ppm	200–3000	150	0
4	Total suspended solids (SS) as ppm	40–500	0	0
5	Anionic surfactant as ppm	0–15	0	0
6	Nonionic surfactant as ppm	2–350	2	<1
7	Cationic surfactant as ppm	2–13	0	0

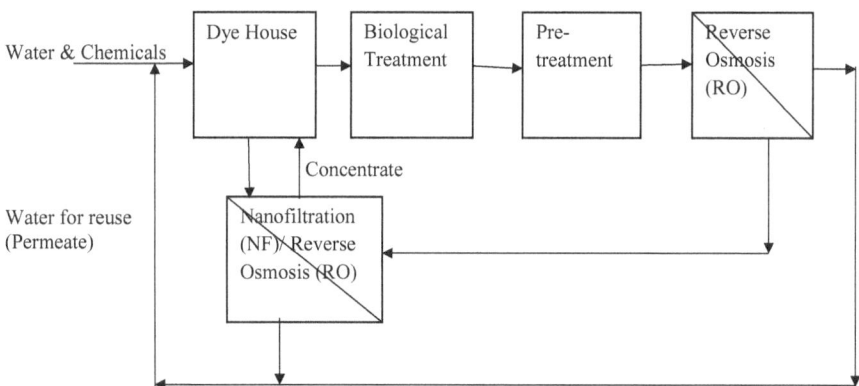

FIGURE 7.5 Recovery and recycling of good-quality water and upgrading of biological water effluent treatment plants in textile industry.

7.5 BENEFITS OF WATER REUSE AND RECOVERY OF VALUABLE MATERIALS

As shown in Figure 7.6, water recycling and reuse has several benefits, including the following:

- Assured availability of quality water for process needs as well as other uses
- Reduced fresh water requirement

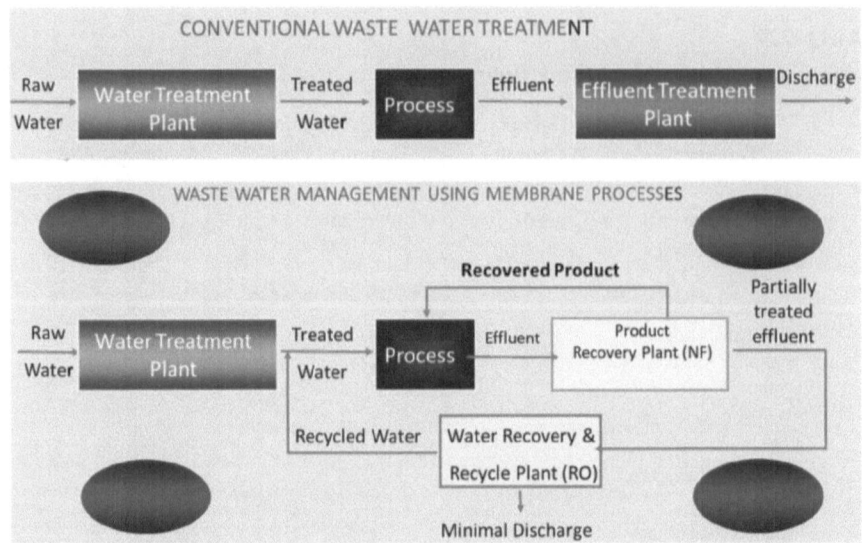

FIGURE 7.6 Wastewater management using membrane processes.

- Possibility of additional savings through recovery of valuable materials from waste streams
- Compliance with pollution control regulations
- Clean environment through reduced effluent discharge
- For new projects, incorporation of recycling may considerably reduce capital expenditure on water treatment
- Recharging of groundwater with cleaner water
- Wetland development

Membrane separation plays an important role in wastewater treatment because it offers a wide choice of separation processes ranging from suspended solids to dissolved ions. Membrane systems are modular in nature. They are ambient temperature operations requiring minimal use of chemicals. They can be adapted to point-of-use applications and support treatment, recycling and reuse, thereby improving the efficiency and economics of wastewater treatment technologies.[10–12]

REFERENCES

1. Yokomizo, T. 1994. Ultrafiltration membrane technology for regeneration of building wastewater for reuse. *Desalination* 98: 319–326.
2. Zhou, Q., et al. 1996. An integrated plan for town-enterprise wastewater reuse and wetland strategy: A case study. *Desalination* 106: 439–442.
3. Wen, S., et al. 1985. *Environmental Microbiology*. Beijing: Academic Press.
4. Dec, J. and Bollog, J.M. 1994. Use of plant material for the decontamination of waters polluted with phenols. *Biotechnology and Bioengineering* 44: 1132–1139.

5. Anderson, T.A., et al. 1993. Bioremediation in the rhizosphere. *Environmental Science and Technology* 27: 2630–2636.

6. Rautenbach, R. and Albrecht, R. 1989. *Membrane Processes*, Chapters 7–10. Chicester, UK: John Wiley & Sons.

7. Kolega, M., Grohsman, G.S., Chiew, R.F. and Day, A.W. 1991. Disinfection and clarification of treated sewage by advanced microfiltration. *Water Science and Technology* 23: 1609–1618.

8. Fabiani, C., Ruscio, M., Spadoni, M. and Pizzichini, M. 1996. Chromium (III) salts recovery process from tannery wastewaters. *Desalination* 108: 183–191.

9. Gaeta, S.N., Fedele, U. and Vaglio, M. 1991. Energy recovery from liquid effluents of the textile industry by membrane processes combined with electric and thermal energy consumption. *European Seminar on New Technologies for the Rationale Use of Energy in the Textile Industry in Europe*. CEE Dir. XVII, Thermie Programme, Milan, Italy.

10. Sarkar, P., Prabhakar, S., Tiwari, S., Goswami, D. and Tewari, P.K. 2011. Recovery of water from saturated solutions by membrane processes. *Desalination and Water Treatment* 36: 65–74.

11. Satpati, S.K., Pal, S., Goswami, D., Tewari, P.K. and Roy, S.B. 2015. Extraction of uranium from nuclear industrial effluent using polyacrylhydroxamic acid sorbent. *International Journal of Environmental Science and Technology* 12: 255–262.

12. Ghosh, A.K., Bindal, R.C., Prabhakar, S. and Tewari, P.K. 2013. Concentration of ammonium diuranate effluent by reverse osmosis and forward osmosis membrane processes. *Desalination and Water Treatment* 52: 1–3.

8 Guidelines for Setting up Advanced Water Technology Plant

8.1 INTRODUCTION

Water issues currently faced by many countries have increased interest in developing and deploying advanced water technologies. However, more effort is required if such projects are to effectively penetrate the market. Many elements are involved, including the business model, various stakeholders and the relationship between vendors and users. An improved understanding of the requirements and constraints of users and vendors, assisted by preliminary exchange of information, is important to facilitate effective implementation of advanced water technology projects. More importantly, vendors and users need to agree on the responsibilities of each party throughout the project. Typically these concern siting, target utilization of the cogenerated products, status and selection of technologies, financial constraints and social responsibility in the form of a contribution towards solving water issues, climate change mitigation and sustainable development. Depending on the type of feed water—raw water, seawater or effluent—and the end use of the product water, various advanced technologies and strategies can be used to produce, recycle or reuse water as well as recover valuable materials from the waste stream.

Several business models can be proposed for advanced water technology projects because, unlike conventional water treatment plants, a number of different products such as different qualities of water or valuable materials may be produced. Depending on the extent of integration among stakeholders (owners, operators, users, etc.), additional user and vendor requirements and responsibilities may be involved. There are differences in design and implementation between retrofitted and new-build advanced water technology projects.

There are certain preliminary requirements for introducing advanced technologies for desalination and water reuse in any country. These include the development of environmental policies and regulations, feasibility studies, public consultations, technology evaluation, requests for proposals, evaluation of proposals, project and contract development, financing, supply, construction and commissioning, and finally operation. The development and implementation of an appropriate infrastructure to support advanced water technologies, as well as their safe and efficient operation, is important, especially for countries that are planning their first advanced water treatment plant. Advanced water technologies of appropriate quality, efficient operation, economic competitiveness, resource utilization, sustainability and better environmental impact should be introduced to fulfill or augment water requirements. The efficient management of project activities is quite demanding and presents a major challenge for utility, regulatory, supplier and other support organizations. The main objective should be to ensure that the project is implemented successfully from a commercial point of view while complying with the appropriate engineering and quality requirements and safety standards.

Some of the requirements for users interested in setting up an advanced water treatment system are:

   i. Selection of the appropriate water technology and plant configuration by assessing criteria specific to the plant including feed analysis, product water quality requirements, etc.
  ii. Assessment of the land requirement, preparation of general layout and process flow diagram.
 iii. Estimation of auxiliary services required, such as power and other utilities for the plant.
  iv. Preparation of preliminary estimates of capital and operating cost.
   v. Preparation of project implementation schedule.

8.2 TECHNOLOGY SELECTION

The energy requirements, capital cost, operating cost and reliability are the key parameters for selection of appropriate advanced water technologies. The choice will depend on location, and on specific conditions prevailing at the site, such as existing facilities, power availability, land availability, fluctuation in feed water conditions, plant capacity, feed water quality, quality of water to be produced, work required by local authorities, etc. Conventional water treatment technologies have been widely used worldwide for treatment and purification of raw water and wastewater. Membrane-based technologies and other advanced water technologies have emerged for desalination, water recycling and reuse. As the technologies are energy intensive, the availability of cheap energy or waste heat can be a deciding factor in process selection.

Many recent and important improvements in membrane-based systems have helped increase their potential in the field. The total installed capacity of membrane technology continues to increase. The performance of membrane materials and modules has improved with respect to salt rejection, higher surface area per unit volume, flux enhancement, improved membrane life and capacity to work at higher pressure at competitive membrane cost. The recovery ratio has increased considerably over the years due to improved salt rejection. The recovery ratio for normal seawater desalination (35000 mg/liter of salinity) was about 25% in the 1980s, increasing to 35% in the 1990s. Currently, it is about 45% and can reach 60% if the second stage is applied. Improved recovery has facilitated a reduction in investment and operating costs. Energy is recovered from the brine side of the reverse osmosis (RO) process either through turbo-systems that include reversible pumps, Pelton turbines, turbochargers and a hydraulic pressure booster, or through volumetric systems that include a pressure exchanger. With the new energy recovery systems that have been developed over the years, the objective of reducing energy consumption below 3 kWh/m^3 for the RO section has been achieved.

The chosen water technologies, basic plant design, capital expenditure (CAPEX) and operating expenditure (OPEX) are planned in detail based on field studies and technical evaluations. The plant includes the feed water system, reject disposal, post-treatment, chemical dosing, storage system, plant electrics, instrumentation and automation. The envisaged plant should preferably be based on proven advanced technology, should reflect state-of-the-art innovation in its design, operation and maintenance, and should include multidisciplinary contributions based on experiences acquired in advanced water technologies.

8.3   SITE SELECTION AND FIELD STUDIES

The location of the plant is decided by the users. A topographical map of the proposed site showing latitude and longitude of its four corners is quite useful. The site should be well connected by road and other means of transport. A connectivity map and information about the nearest human settlement is important. Topographical and geographical details are provided by the user to the vendor for plant construction purposes, together with information about availability of power and raw water sources. The proposed site should not be located in reserve forests or forest land, in the vicinity of any monument or in an archeologically sensitive area, and there should be no biodiversity parks/sanctuaries in the surrounding area. A techno-economic feasibility report (TEFR) should be prepared based on the field studies and existing multidisciplinary experience of similar plants.

In coastal areas, comprehensive marine studies, including bathymetry, sub-bottom profile, side-scan surveys and seismic studies, provide data on the nature of the sea, the sea-bed profile with respect to distance from shore, high and low tide levels, wave patterns and seawater quality including fouling characteristics. For example, bathymetric studies play an important role in the design of the seawater intake and outfall system. It is preferable to locate the plant relatively close to the point where the seawater can be introduced into the system. This reduces the extent of pretreatment required, reducing operating cost and operational risks. To maintain reasonable inlet water quality, it is preferable to have the inlet located in deeper water. This avoids sediments that are stirred up by wind and wave action in shallower waters. It is also preferable to keep the inlet sufficiently deep to avoid any interaction with boats and shipping. Further, constructing connecting pipes or tunnels to the inlet is often costly, so minimizing this length is beneficial. This means that sites where deeper water is relatively close to shore are preferred. The concentrate or effluent can be dispersed effectively using diffusers on the outlet pipeline, and since this requires a minimum depth of water, again sites near deeper water are preferred. The seawater or effluent outfall should be separated from the intake point sufficiently to avoid short circuiting. In the case of a desalination plant, the seawater from the intake chamber is pumped to the plant. The seawater requirement for a desalination plant depends on the design aspects and technology used. A thermal desalination plant such as MSF or MED requires seawater for condenser cooling as well as feed. An RO plant only requires seawater as feed. The pretreatment system for feed seawater is different in both cases. Pretreatment for a thermal desalination plant is simple as it can tolerate a higher contamination level in seawater than an RO plant. The advanced pretreatment system for a typical seawater RO desalination plant may comprise a back-washable disc filter, ultrafiltration system, etc. Conventional pretreatment equipment like clarifiers, tube settlers, gravity and pressure sand filters is also used instead of membrane-based UF pretreatment. Data on the annual variation in salinity and quality of the seawater in the proposed location is important for system designers as well as users and vendors. It helps to decide the equipment rating and recovery for the salinity range. Users are required to establish the seismic potential of the site by reference to local seismic zone data. Information on site altitude should also be available to ensure it is at a safe elevation above mean sea level (MSL).

8.4   UTILITIES AND AUXILIARY SERVICES

It is important to be clear about power availability. An instrumentation system that has adequate measurement and control facilities is important for achieving efficient, reliable and trouble-free operation, as well as safety of the plant, equipment and operating personnel and user-friendly man–machine communication.

A systematic approach to the layout should be adopted, ensuring that all subsystems are optimally laid out from the viewpoint of operational and maintenance flexibility, and to optimize land use, and capital and operating costs. Approach roads should be planned between the main building and the plant. In particular, the chemical building, bulk chemical storage and handling area, electrical and control room building, and post-treatment area should be easily approachable by road. Within each facility, the user determines the approach to various parts of the equipment. Detailed descriptions of the key components of the proposed plant should be available.

Users must understand the details of the process. For example, in the RO desalination plant, seawater is received through a pipeline into an onshore intake chamber from where it is pumped to the automatic back-washable disc filter and ultrafiltration system for removal of suspended solids, including oil and grease. The feed seawater then is pumped at high pressure through the RO membrane system for separation of dissolved solids, producing desalinated water. The dissolved solids remain in the reject stream, which imparts its energy to the RO feed water in the energy recovery device. The reject brine stream from the energy recovery device is disposed of into the sea through a reject disposal pipeline and diffuser system. The permeate water from the RO system is subjected to recarbonation in the post-treatment section to produce potable water.

Storage, handling, preparation and dosing facilities for chemicals like ferric chloride, polyelectrolytes, sodium bisulfate, sodium hypochlorite and limestone should be provided. A centralized chemical preparation and dosing system required at various process stages should also be considered. Auxiliary plant facilities, including power supply and distribution, electrical system, instrumentation and automation, air conditioning and ventilation, material handling and laboratory facilities, should be envisaged by the user. Administrative buildings should also be provided for as part of the plant facilities. Seawater is corrosive and plants have both low-pressure and high-pressure requirements. Construction material suitable for low-pressure and high-pressure equipment, pumps, piping, valves and fittings should be specified by the user.

The major components and units in a typical RO seawater desalination plant are as follows:

i. Sea water intake and outfall facilities
ii. Water conditioning by chemical dosing
iii. Pretreatment by back-washable disc filter followed by UF membranes
iv. Desalting by RO membrane
v. Energy recovery system
vi. Intermediate storage tanks
vii. Storage, dosing and pumping system
viii. Recarbonation and remineralization system
ix. Electrical power distribution
x. Instrumentation and control
xi. Associated facilities

8.5 PROJECT DESCRIPTION

The capacity of the plant and the qualities of the product water and valuable materials to be produced are decided by the user based on the various end uses. The project description includes details of the systems and subsystems, including the latest innovations. The monthly consumption of chemicals by the proposed cogeneration plant should be estimated. The raw water or seawater requirement and its source should be worked out. The indicative power requirement per day, including connected load and operating load, should be estimated. The type of contract and the estimated manpower requirement in different phases should be decided. It is important to demonstrate that due consideration has been given to making the environmental impact positive, neutral or, if adverse, minimal.

8.5.1 PROCESS DESCRIPTION

The process description for a typical seawater desalination plant may include that it envisages a seawater intake system with stop log gates, trash racks and traveling water screen arrangement, and an intake chamber with pump house located near the sea shore. The water will be pumped to the pretreatment section using a vertical turbine pump. A double set (one working and one standby) of parallel chambers will be provided before the intake sump, each designed to cope with 100% of the intake flow. Stop log gates, trash racks and traveling water screens are installed in each of the parallel chambers. The feed water salinity, suspended solids load and organics determine the type of pretreatment and desalination systems used. The pretreatment system in an RO desalination plant includes facilities to handle oil content in the feed water, such as a dissolved air flotation system to separate floating and settleable suspended solids present in the sea in order to provide consistent quality to downstream pretreatment units consisting of high-efficiency micron filters and ultrafiltration units for conditioning the feed water. The system includes back-washing carried out with backwash pumps. The RO system uses a seawater RO membrane to produce desalinated water. The RO section also has a chemical cleaning system, which includes chemical preparation tanks and transfer pumps. The post-treatment system comprises equipment for recarbonation and remineralization (if required) of the permeate obtained from the RO section. The recarbonation and remineralization systems treat the permeate to achieve the desired potable water alkalinity and hardness, and a positive Langelier saturation index (LSI) suitable for transportation and distribution. The reject outfall pumping and pipeline system with diffuser are designed and located appropriately.

8.5.2 WASTE GENERATION

The types of waste generated during plant operation must be specified. The following types of waste and effluent are generated by a seawater desalination plant.

8.5.2.1 Brine

The concentrate brine or effluent generated by the desalination plant is pumped back into the sea. The concentration and temperature of brine is maintained in accordance with environmental norms.

8.5.2.2 Sewage

The plant will be operated and maintained by the staff. The sewage generated from the proposed plant will be treated in the sewage treatment plant.

8.5.2.3 Solid Waste

About 15–20% of membrane elements are replaced every year in an RO plant. Membrane elements are made up of polymers such as polyamide in the case of RO and polysulfone in the case of UF. Used membrane elements are disposed of appropriately.

8.6 SAFETY CONSIDERATIONS

The basic safety considerations include:

- Adopting safe and quality construction practices
- Adopting safe operation and maintenance practices
- Adopting proven designs for plant, process and equipment

The main objective of safety considerations is early detection and prevention of accidents and, if they do occur, using engineered features to contain the damage within the site, safeguarding operating personnel and the public. To achieve this, a defense-in-depth approach is adopted. Facilities coming into the category of advanced water technologies are connected with chemical plants, but only handle large quantities of water. Inadvertent critical events like chemical explosions, fire involving materials or storage tank breaches are not likely to occur, since they are not involved in the process. The probability of such events is almost nil. However, chemicals used in pretreatment and conditioning of the raw or effluent water may cause accidents if not properly handled. These accidents can be controlled by (i) safe design geometry of tanks and equipment, (ii) limiting the concentration and (iii) effective administrative control of plant operation and relevant equipment. A safety review committee may be formed to monitor and regulate the safety aspects of the plant.

8.7   PROJECT COMPLETION PERIOD

Users must work out the estimated time to complete the project, including construction, testing, commissioning and performance guarantee test period.

8.8   CONTRACT AND DELIVERY MODELS

Several contract and delivery models are available for project delivery. Each has its own strengths and weaknesses. Some of the popular contract and delivery models are as follows:

- Turnkey basis
- Split-package basis
- Engineering, procurement and construction (EPC)
- Design–build (DB)
- Design–Build–Operate (DBO) and maintain (DBOM)
- Design–Build–Own–Operate–Transfer (DBOOT)
- Build–Own–Operate–Transfer (BOOT)
- Design–Build–Own–Operate (DBOO)
- Build–Own–Operate (BOO)

Vendor and user responsibilities will depend on the type of contract. In a turnkey project, the vendor is responsible for the whole project, based on the user requirements, with minimal responsibility for the user. The turnkey approach is preferred when users have little project management and heavy construction experience and would require lengthy extensive training to attain the necessary skills. In a split-package project, the responsibilities are divided, with the user typically taking responsibility for overall project management and decision making and the vendor responsibilities being divided among one or more contractors and limited to the scope of their individual work. The major benefit of a split-package project is that the cost to the user is lower than in a turnkey project. However, a well-qualified and experienced user is required for split package-type projects.

The EPC contract and delivery model is also quite popular. Some industries prefer to follow the DBOM or EPC model with an operation and maintenance contract of up to seven years' duration.

In all these cases, users must clearly define the scope of the vendor's work during site development, construction, installation, testing and commissioning. Scope with respect to power supply and utilities such as water and air should also be clearly defined for the whole term of the project.

8.9   FINANCIAL ANALYSIS

Cost estimation of the project is dependent on local factors, the end user, the economics of the plant's capacity, product water quality, redundancy, type of tanks and buffer storage, location-specific intake and outfall system, prevailing tax structures, etc., and cannot therefore normally be compared with similar projects of the same capacity elsewhere. Normally, the higher the plant capacity, the lower the product cost and investment cost per cubic meter of product. At times, political or environmental issues can also be a limitation on the successful implementation of megaprojects.

Capital and operating costs generally depend primarily on the following aspects:

- Geographical location and site-specific characteristics.
- Feed water quality, temperature, intake arrangements and required product water quality.
- Reject discharge type and product water storage.
- Materials, equipment, chemicals and other consumables.
- Financing details and amortization period as well as inflation.
- Production cost depends on the cost of materials, consumables, labor and supervision, power and other services.

New environmental regulations encourage designers and plant owners to develop advanced methods for effluent, particularly concentrate, discharge. Concentrate salinity and chemistry, temperature and hazardous chemicals are the main concerns. Depending on the reliability of the treatment plant, the infrastructure and electric power source, and the need for emergency storage of produced water, on-site storage capacity can range from a few hours to a few days. The energy cost is normally included in the contract agreement for the service period as part of the total water cost. However, virtually all BOO and BOOT contracts contain an energy adjustment cost provision to cover variations in electric power costs. Therefore, minimization or reduction of the facility's energy consumption has a major impact on reducing the unit water cost. A financial analysis of capital and production costs should be carried out for the proposed plant.

8.10   ENVIRONMENTAL IMPACT ASSESSMENT

Advanced water technologies have the potential to produce, recover and reuse water for the sustainable development of society. The actual impact of a project can be gauged by comparing the environmental parameters at a later date with their current values. This calls for the generation and establishment of baseline data, and the institution of an effective environmental monitoring program for the future. The likely benefits, stresses and loads on the environment due to the expected activities of the project should be identified.

Baseline data are classified according to their application. The categories are:

- parameters indicating environmental status/quality
- data related to use of natural and man-made resources
- meteorological data used as inputs to models for impact prediction and site evaluation
- marine data used as inputs to model for brine dispersion

In many cases, setting up the plant requires prior environmental clearance. If it is situated in a coastal area, the approval of the Coastal Zone Management Authority and Coastal Regulation Zone clearance are needed. In order to obtain environmental clearance, an environmental impact assessment (EIA) study is carried out in and around the proposed project site. The report should include details of the measures to be taken to eliminate or minimize adverse impacts from the siting, construction and operation of the proposed nuclear desalination plant. Positive environmental impacts should also be emphasized.

8.10.1 SCOPE OF THE EIA STUDY

For the EIA study, users should finalize the capacity and size of the advanced water treatment plant. The scope of the project and the mode of its execution, including infrastructure, design, detailed engineering, systems and subsystems procurement, erection, testing and commissioning, should be decided. The scope of work required for site-related activities like leveling, infrastructure construction, electrical work, intake, reject disposal, utilities and statutory clearances should be confirmed.

The EIA study consists of the following elements:

I. Generate (a) baseline information regarding environmental status through baseline survey of environmental parameters like air, water, noise and soil; and (b) secondary data required for impact assessment from available sources.

II. Identify activities that have a potential impact on the environment or the public and provide mitigation measures having regard to project details.

III. In the case of coastal sites, prepare a Coastal Regulation Zone map showing CRZI, II, III and IV areas, including other notified ecologically sensitive areas and marking high tide line (HTL), low tide line (LTL) etc. Carry out marine ecological studies to find out the quality of seawater and impact of brine disposal, impingement and entrainment of marine organisms near proposed intake and outfall diffuser point. Conduct bathymetry, side-scan and seismic surveys to measure current, tides, waves and sea-bed sediment quality, and to profile the sea bed for intake channel and outfall diffuser design.

IV. Carry out land-use and land-cover (LULC) pattern studies for the proposed project site and study the area to identify impacts on land.

V. Predict pollution load/stress level in air, water, land and other environmental matrices from the sources identified above.

VI. Examine availability and adequacy of the envisaged measures for control of pollution and sources of stress, so as to meet statutory provisions, and propose mitigating measures, if needed.

VII. Analyze the consequences of potential accidents, steps adopted to avert accidents and plans to mitigate consequences of severe accidents.

VIII. Specify environmental monitoring required in the operational phase of the plant to evaluate the effectiveness of the various environmental control measures adopted.

IX. Assess the benefits arising from the project.

X. Identify administrative mechanisms warranted to oversee (a) implementation of control, environment management plan, mitigating measures before commissioning of the plant, (b) subsequent operation and maintenance of such systems, (c) compliance with monitoring programs and (d) provision of required budget.

There are significant gaps in understanding between the user and vendor when setting up a project based on advanced water technology compared to a conventional water treatment project. A generic algorithm needs to be developed to define the roles of various stakeholders in general, and users and vendors in particular, taking account of the technology, business models, regulations, public sensitivity, media involvement and scientific groups. The responsibilities of vendor and user depend on the type of project, type of technology, retrofit or new build, and type of contract. The requirements and responsibilities of user and vendor at various stages of the project life cycle should be addressed. Development of guidelines for vendors and users of retrofit and new-build projects covering the front-end cycle for production and purification of water for different applications, and the back-end cycle for effluent treatment, recycling and reuse, and recovery of valuable materials, is extremely important.

General technical information on commercial water technologies is given in Tables 8.1–8.13.

TABLE 8.1
Multi-Stage Flash (MSF) Technology for Seawater Desalination

S. No.	Items	General Technical Information
1	Objective of the technology	To produce distilled-quality water from seawater
2	Features of the technology	• Uses low-pressure steam • Specific energy consumption: Less than 2 kWh/m^3 and steam 0.1 tonne/hour (at 2 bar pressure) per cubic meter of desalinated water
3	Capacity	Large capacity
4	Indicative capital cost	About US$ 1 million per MLD capacity at battery limits
5	Indicative O&M cost	US$ 0.5 per cubic meter of desalted water
6	Indicative space requirement	1000 square meters for 1 MLD capacity plant

TABLE 8.2
Multi-Effect Distillation-Mechanical Vapor Compression (MED-MVC) Technology for Seawater Desalination

S. No.	Items	General Technical Information
1	Objective of the technology	To produce distilled-quality water from seawater
2	Features of the technology	• Uses electricity alone • Uses high-speed mechanical vapor compressor
3	Capacity	Small to medium capacity
4	Indicative capital cost	About US$ 1.5 million per MLD capacity at battery limits
5	Indicative O&M cost	US$ 0.7 per cubic meter of desalted water
6	Indicative space requirement	800 square meter for 1 MLD capacity plant

TABLE 8.3

Multi-Effect Distillation-Thermal Vapor Compression (MED-TVC) Technology for Seawater Desalination

S. No.	Items	General Technical Information
1	Objective of the technology	To produce distilled-quality water from seawater
2	Features of the technology	• Uses steam • High efficiency and high gain output ratio • Reduced condenser cooling water requirement
3	Capacity	Medium capacity
4	Indicative capital cost	About US$ 0.8 million per MLD capacity at battery limits
5	Indicative O&M cost	US$ 0.5 per cubic meter of desalted water
6	Indicative space requirement	1000 square meter for 1 MLD capacity plant

TABLE 8.4

Low-Temperature Evaporation (LTE) Technology for Seawater Desalination Utilizing Waste Heat

S. No.	Items	General Technical Information
1	Objective of the technology	To produce distilled-quality water from seawater
2	Features of the technology	Uses waste heat in form of hot water/low quality steam as energy source for producing distilled-quality water
3	Capacity	Small capacity
4	Indicative capital cost	About US$ 1.5–2 million per MLD capacity at battery limits
5	Indicative O&M cost	US$ 0.2–0.3 per cubic meter of desalted water
6	Indicative space requirement	1000 square meter for 1 MLD capacity plant

TABLE 8.5

Back-Washable Spiral Wound Ultrafiltration (UF) Element

S. No.	Items	General Technical Information
1	Objective of the technology	Purification of water for the removal of suspended/colloidal and biological contaminants
2	Features of the technology	• For community or industrial applications • Can replace conventional filtration systems like sand filter and cartridge filter • Back washable in auto/manual mode
3	Capacity	Modular in nature Single-unit capacity varies from 1000 to 7000 LPD depending on element.sizes
4	Indicative capital cost	About US$ 0.1 million per MLD capacity at battery limits
5	Indicative O&M cost	US$ 0.1 per cubic meter of purified water
6	Indicative space requirement	200 square meter for 1 MLD capacity plant

TABLE 8.6
Candle-Type UF Membrane Device

S. No.	Items	General Technical Information
1	Objective of the technology	Removal of turbidity, colloids and microbiological contaminants
2	Features of the technology	• Suitable for domestic and community applications • Operates under tap water head (3 meters head) • Typical filter life: 3–5 years
3	Capacity	Small size Modular in nature Typical single-unit capacity: 100 LPD
4	Typical unit cost	US$ 50–100
5	Indicative O&M cost	Negligible (no electricity or chemicals required)
6	Indicative space requirement	Small units up to 100 LPD can be installed as a wall-mounted unit. Bigger units may require about 2.0 square meters of footprint area excluding storage tank

TABLE 8.7
Fluoride Removal Technology for Water Purification

S. No.	Items	General Technical Information
1	Objective of the technology	Treatment of fluoride-contaminated groundwater to obtain safe drinking water as per standard
2	Features of the technology	• Ultrafiltration (UF) membrane-assisted alumina adsorption process • Fluoride ion in feed that can be treated: 10 ppm (max) • Product water free from aluminum (less than 0.1 ppm), biological and colloidal contaminants
3	Capacity	100–5000 liters per day per unit
4	Typical unit cost	US$ 50 and above
5	Indicative O&M cost	US$ 0.1 per cubic meter of purified water
6	Indicative space requirement	Small units up to 100 LPD can be installed as a wall-mounted unit. Bigger units may require about 2.0 square meters of footprint area excluding storage tank

TABLE 8.8
Arsenic Removal Technology for Water Purification

S. No.	Items	General Technical Information
1	Objective of the technology	Treatment of arsenic-contaminated groundwater to obtain safe drinking water as per standard
2	Features of the technology	• Ultrafiltration (UF) membrane-assisted sorption process • Arsenic ion in feed that can be treated: 500 ppb (max) • Product water contains less than 10 ppb arsenic concentration • Product water free from biological and colloidal contaminants
3	Capacity	100–5000 liters per day per unit
4	Typical unit cost	US$ 50 and above
5	Indicative O&M cost	US$ 0.1 per cubic meter of purified water
6	Indicative space requirement	Small units up to 100 LPD can be installed as a wall-mounted unit. Bigger units may require about 2.0 square meters of footprint area excluding storage tanks

TABLE 8.9
Iron Removal Technology for Water Purification

S. No.	Items	General Technical Information
1	Objective of the technology	Treatment of iron-contaminated groundwater to obtain safe drinking water as per standard
2	Features of the technology	• Ultrafiltration (UF) membrane-assisted oxidation process • Iron (ferrous ion) in feed that can be treated: 20 ppm (max) • Purified water quality: Iron below 0.1 ppm • UF-filtered product water is also free from biological and colloidal contaminants
3	Capacity	100–5000 liters per day per unit
4	Typical unit cost	US$ 50 and above
5	Indicative O&M cost	US$ 0.1 per cubic meter of purified water
6	Indicative space requirement	Small units up to 100 LPD can be installed as a wall-mounted unit. Bigger units may require about 2.0 square meters of footprint area excluding storage tanks

TABLE 8.10
Brackish Water Reverse Osmosis (BWRO) Desalination Technology

S. No.	Items	General Technical Information
1	Objective of the technology	To desalinate brackish water so as to produce good water for domestic or industrial uses
2	Features of the technology	• Uses RO technology • UF pretreatment (if required) • Product water post-treatment (if needed) • Reject management of harmful contaminants for disposal as per environmental norms • Site-specific design
3	Capacity	Small and medium size
4	Indicative capital cost	US$ 0.3–0.7 million for 1 MLD capacity
5	Indicative O&M cost	US$ 0.2–0.5 per cubic meter of purified water
6	Indicative space requirement	About 2.0 square meters of footprint area for 1 KLD capacity

TABLE 8.11
Seawater Reverse Osmosis (SWRO) Technology

S. No.	Items	General Technical Information
1	Objective of the technology	To desalinate seawater so as to produce good water for domestic or industrial uses
2	Features of the technology	• Uses RO technology • UF pretreatment (if required) • Product water post-treatment (if needed) • Reject management of harmful contaminants for disposal as per environmental norms • Site-specific design
3	Capacity	Small to large size
4	Indicative capital cost	US$ 0.8 million for 1 MLD capacity
5	Indicative O&M cost	US$ 0.5 per cubic meter of purified water
6	Indicative space requirement	2000 square meters for 1 MLD capacity plant

TABLE 8.12
Small-Size Solar Reverse Osmosis (RO) Unit for Brackish Water Desalination

S. No.	Items	Specific Information
1	Objective of the technology	To desalinate brackish water (1000–2500 ppm) so as to produce drinking water (100–200 ppm)
2	Name of the process	Solar reverse osmosis (RO)
3	Features of the technology	• Conserves groundwater sources • Cartridge pre-filter • Spiral RO membrane element
4	Capacity	100 LPD
5	Approximate capital cost	Depends on feed water quality and local factors
6	O&M cost	Depends on local conditions
7	Specific prerequisites for installation of the unit	
	a) Open space/shed required for installation of the unit	Domestic use
	b) Electrical power requirement	Nil
	c) Specialized manpower requirement	Nil

TABLE 8.13
Community-Size Solar Reverse Osmosis (RO) Plant for Brackish Water Desalination

S. No.	Items	Specific Information
1	Objective of the technology	To desalinate brackish water (2000–2500 ppm) so as to produce drinking water (200–250 ppm)
2	Name of the process	Solar reverse osmosis (RO)
3	Features of the technology	• Conserves groundwater sources • Pre-filters • Product post-treatment for palatability
4	Capacity	2 KLD
5	Approximate capital cost	Depends on local factors
6	O&M cost	Will vary depending on local conditions
7	Specific prerequisites for installation of the unit	
	a) Open space/shed required for installation of the unit	About 20 square meters for 2 KLD
	b) Electrical power requirement	Nil
	c) Specialized manpower requirement	Local manpower with minor training

9 Challenges and Opportunities

9.1 INTRODUCTION

Advanced water technologies are a boon, particularly for water-starved locations. Capacity has rapidly increased because of increased water demand and advances in water technologies. Membrane science has shown remarkable developments in the field of desalination and water reuse technologies, leading to greater reliability and affordability. Through modular units, the benefits of technology have been extended to remote locations requiring small and medium capacities at reasonable cost. With increasing environmental pollution and dwindling natural sources of uncontaminated fresh water, desalination and water purification has become an essential component in the integrated management of water resources, even for water-rich countries. Hence, it is necessary to focus on ways of further reducing the cost of desalinated water, ensuring safety, reliability, flexibility and environment friendliness. There is a need to enhance interdisciplinary expertise ranging from membrane development, advanced energy recovery devices and improved construction materials to new design schemes and use of renewable energies. This means converting concepts into technologies and fine tuning existing technologies to develop cheaper components, reduce maintenance costs, increase system life and add value through smart management of concentrate streams. R&D inputs are required to develop eco-friendly, simple and efficient new concepts that utilize new and renewable energy sources. Furthermore, since water is a societal need, provision may also be needed to ensure its availability at affordable cost to resource-poor users.

Plant suppliers and consultants have their own cost evaluation methodologies and calculate water costs with different degrees of accuracy.[1] Existing cost evaluation methodologies and software packages such as WTCost© and DEEP provide indicative estimates and their applications are limited to specific conditions. Generally, available cost-estimation tools provide few details of their parameters and methodologies. A generalized cost-estimation tool helps in selecting an appropriate desalination technology suitable for a specific location taking all the relevant parameters into account. Technical developments, such as the increase of unit capacity, improvement in process design and materials, and hybrid systems, have contributed to cost reduction through technological innovations. The development of new and emerging low-energy desalination technologies will have a further positive impact on the techno-economic aspects of these innovations.

9.2 CHALLENGES

Research and development in water technologies is more challenging than research in some other fields in view of its societal aspects. Water is required by all sections of society from resource-rich to resource-poor users, including those at the bottom of the pyramid in rural and remote areas. An ongoing focus is needed, not only on continuous advancement but also on the application of research findings to improving cost effectiveness and performance, while also meeting environmental challenges.

9.2.1 ENVIRONMENTAL CHALLENGES

Desalination and water reuse technologies ranging in size from large to small systems produce purified water from saline water and wastewater using energy. They discharge concentrate stream to the environment. Comprehensive studies on the environmental impact of emissions from desalination and water reuse units are important.

9.2.1.1 Environmental Considerations for Seawater Desalination Plants

From an environmental perspective, the effect of brine discharged to the sea from seawater desalination plants must be considered carefully on a case-by-case basis with respect to the following constituents:

   (i) Corrosion products
  (ii) Antiscalant additives
 (iii) Anti-foaming agents
  (iv) Antifouling additives
   (v) Flocculants
  (vi) Halogenated organic compounds after chlorination
 (vii) Anti-corrosion additives
(viii) Oxygen scavengers
  (ix) Acids
   (x) Concentrate discharge
  (xi) Thermal pollution

Environmental assessment and monitoring is essential to minimize the impact on marine and terrestrial ecosystems. The extent of the impact may vary in different areas according to the hydro-geological nature of the marine body as well as the type and size of desalination plant, the required secondary structures and infrastructure. Higher salinity of brine discharge can damage the cell walls of marine organisms, causing dehydration. Noise pollution from pumps should also be abated by acoustic means. An environmental impact assessment regarding air, land (surface and underground) and sea is desirable. Desalination initiatives sometimes face resistance from local fishermen apprehensive about the ill effects of concentrated brine discharge into the sea.

There are different types of sea water intake systems for desalination plants. Selection of a particular type of intake system is based on various criteria, each with its own merits and demerits. For example, a beach well-type intake system gives filtered raw water of stable temperature requiring less pretreatment, but may adversely affect the water table.

Brine discharge from a membrane-based desalination plant is high in salinity, while brine discharge from a thermal desalination plant is at relatively elevated temperature. Hence it is desirable to assess the thermal and chemical impact on the seawater. Marine dilution by jet mixing and turbulence is helpful in preventing higher-density brine from settling and is useful for spreading on the marine floor. In the case of cogeneration plants, hot water discharged from the power station is also sometimes considered for dilution of the concentrated brine. Mixing the warm low-density stream with cold brine can promote dilution and decrease settling to the bottom. Any leaks in seawater intake pipes and concentrated brine discharge need to be detected by sensors and detectors and acted upon to prevent underground contamination. Measures to extend the life of aromatic polyamide membranes and identify suitable means for their eventual disposal are recommended. Alternatives to continuous discharge of pretreatment chemicals may involve pressure-driven membrane processes such as microfiltration, ultrafiltration and nanofiltration to screen out bacteria, viruses, colloids, suspended matter and scale-forming compounds.

Environmental impact assessment, bathymetric studies, seawater current studies and dispersion models are essential for the design of suitable seawater intake and brine discharge system to meet environmental norms and regulations. If seawater is contaminated, the extensive pretreatment requirements of membrane desalination can lead to generation of hazardous solid waste, which has to be disposed of at suitable landfill sites. Thermal desalination technology has the potential to eliminate such issues. Membrane-based pretreatment is a possible alternative to conventional chemical pretreatment, minimizing chemical requirements. Selection of appropriate pretreatment options may need fewer chemicals and have a less adverse impact on pollution.

Detailed studies are required with respect to the following aspects:

a. Effect on seawater characteristics of continuous discharge of brine with high turbidity and salinity.
b. Discharge of pretreatment chemicals causing environmental pollution.
c. Disposal of used polypropylene cartridge filter elements.
d. Disposal of used thin-film composite (TFC) polyamide membrane elements.

Alternative membrane materials such as cellulose acetate (CA) and cellulose triacetate (CTA) have the potential to deal with the problem of disposal of used membranes because they are eco-friendly. Use of a reusable stainless-steel sintered cartridge filter or ceramic filters may overcome the polypropylene cartridge filter element disposal issues.

Generally, desalination plants have a positive impact as they provide good-quality water, which all living beings need to survive.

9.2.1.2 Environmental Considerations for Brackish Water Desalination Plants

In the case of inland brackish water desalination systems, where brine concentrates are normally disposed of into inland areas, haloculture (high salt cultivation) has been attracting attention and a number of salt-tolerant crops have been used. NyPa forage such as *distichlis spicata* is a typical salt-grass with about 10 times more yield than other euhalophytes (plants having increased productivity with increased salt content). This variety has potential to grow at full strength at seawater concentrations. It is a suitable food for goats, sheep and beef cattle. The Food and Agriculture Organization of the United Nations has software to assist appropriate crop selection for different regions. Serial haloculture is being investigated whereby after each haloculture the drainage water can be used for the next halophytic crop. Since these plants are specific to their location, R&D efforts are required to identify them.

If the brackish water contains contaminants such as fluoride, nitrate, arsenic, etc., the concentrates require isolation from the environment to prevent consumption by animals. Technologies to remove the contaminants from the concentrates in solid form have to be developed and integrated with the units. Thus, the approach to brine disposal should include:

- System integration studies of hybrid desalination plants, their proper sequencing and operation.
- Zero-liquid-discharge desalination systems involving concentrate processing, solids disposal and recovery of evaporated water.
- Treatment of concentrated brine to recover its constituents.
- Developing environmentally sound desalination practices addressing ecosystem impacts such as effects of water withdrawal, water intake structure, etc.

9.2.1.3 Environmental Considerations for Water Purification Plants

Disinfection by-products are formed when disinfectants used in water purification plants react with bromide and/or natural organic matter (i.e., decaying vegetation) present in the source water. Disinfectants produce different types or amounts of disinfection by-products. Trihalomethanes (THMs) are a group of chemicals which are formed along with other disinfection by-products when chlorine or other disinfectants react with naturally occurring organic and inorganic matter in water. Haloacetic acids are formed along with other disinfection by-products when chlorine or other disinfectants used to control microbial contaminants in drinking water react with naturally occurring organic and inorganic matter in water. The regulated haloacetic acids, known as HAA5, are: monochloroacetic acid, dichloroacetic acid, trichloroacetic acid, monobromoacetic acid and dibromoacetic acid. Bromate is formed when ozone used to disinfect drinking water reacts with naturally occurring bromide found in source water. Certain minerals are radioactive and may emit a form of nuclear radiation. Some of the minerals may emit forms of radiation known as beta radiation. People who over the years consume water containing alpha emitters or beta emitters in excess of prescribed limits may be at increased risk of developing cancer. Some people who over the years consume water containing radium 226 or 228 in excess of prescribed limit may also have an increased risk of developing cancer. Drinking water containing arsenic in excess of the standard limits may cause skin damage or problems in the long term, and may increase the risk of cancer. People consuming water containing fluoride levels over the prescribed limit may develop dental fluorosis. Higher fluoride content in water may also cause bone disease, including pain and tenderness.

Table 9.1 gives some of the inorganic contaminants in water, the potential health hazards associated with them and the sources of contamination.

TABLE 9.1

Some Contaminants in Drinking Water, Potential Health Hazards and Sources of Contamination

S. No.	Inorganic Chemicals	Maximum Contamination Level (mg/liter)	Potential Health Hazard	Sources of Contaminant in Drinking Water
1	Arsenic	0.01	Skin damage, circulatory system problems, increased risk of cancer	Erosion from natural deposits, discharge from semiconductor manufacturing, petroleum refining
2	Fluoride	1.5	Dental fluorosis, bone disease	Erosion from natural deposits, discharge from fertilizer and aluminum factories
3	Chromium	0.1	Allergic dermatitis	Erosion from natural deposits, discharge from tannery, steel and pulp industries
4	Lead	–	Kidney problems, high blood pressure	Erosion of natural deposits, industrial effluents
6	Nitrate	10	'Blue baby syndrome' in infants	Run-off from fertilizer use, leaching from septic tanks
7	Selenium	0.05	Hair or fingernail loss	Effluent from petroleum refineries, erosion of natural deposits, industrial effluents
8	Cadmium	0.005	Kidney damage	Erosion of natural deposits, corrosion of galvanized pipes, run-off from waste batteries and paints
9	Total coliforms including fecal coliform and *E. coli*	10	Waterborne diseases	Human and animal fecal waste
10	Viruses	0	Gastroenteric diseases	Human and animal fecal waste
11	Gross alpha particles activity	0.1 Bq/liter	Increased risk of cancer	Decay of natural and manmade deposits
12	Gross beta particles activity	1.0 Bq/liter	Increased risk of cancer	Decay of natural and manmade deposits
13	Radium 226 and Radium 228 (combined)	5 pCi/liter	Increased risk of cancer	Erosion of natural deposits

9.3   RURAL APPLICATIONS OF THE TECHNOLOGY

Desalination is normally resorted to when conventional options for providing drinking water are exhausted. Appropriate methods must be adopted for the disposal of concentrated brine in inland areas. If plants are located in coastal areas, the reject water can be directed towards the sea and disposed of in accordance with environmental norms.

9.4   ENVIRONMENTAL PROTECTION AND POLLUTION PREVENTION

The concentrate from desalination plants is harmful to the environment as it may re-contaminate their sources. Seawater concentrates can be discharged back to the sea in accordance with environmental norms or can be used for the recovery of salts, if conditions permit. In the case of brackish water concentrates, proper methods of brine management based on haloculture/solar evaporation, etc. may be identified and applied.

9.5   OPPORTUNITIES

It is widely recognized that the availability of water for domestic, agricultural and industrial requirements is going to be a serious constraint in several areas of the world, and that this may have an adverse impact on economic development and human health. There is a growing need and opportunity to develop and introduce a science and technology-based water security system that is economically and environmentally sustainable.

The cost of water produced in seawater desalination plants has been falling over the years, whereas the cost of conventional water treatment plants is rising due to the increasing contamination of groundwater, overexploitation of aquifers and the intrusion of saline water in coastal areas. There is plenty of scope and opportunity for the development of desalination and water purification technologies, equipment and components.

Desalination is an attractive alternative in salinity-affected areas and coastal zones, wherever there is acute water shortage. Energy is the major input for water desalination. In view of the limited availability of fossil-fuel sources as well as the environmental pollution associated with conventional energy sources, clean energy sources are a promising prospect. It is anticipated that there will be demand in the market for both large and small desalination plants. Both stationary land-based and mobile desalination units will be required to cater to the needs of different water-scarce coastal areas. Thermal and membrane-based desalination technologies to meet desalinated water requirements will continue to be the front runners commercially. Newer developments such as nanocomposite membranes, hybrid systems, advanced membrane-based pretreatment, energy recovery systems, nuclear desalination, recovery of valuable materials from the reject brine and other technological innovations are among the cost-reduction strategies that have potential to enrich the field.

Significant scope for interdisciplinary research exists, ranging from water quality and wastewater issues (environmental-chemical-civil engineering), minimizing energy consumption (chemistry-chemical-mechanical engineering), energy devices (electrical-mechanical engineering) and membrane development (chemistry-chemical-material engineering and sciences) to innovations in construction materials to handle corrosive water (chemistry-chemical-metallurgical engineering), new design schemes and use of renewable energies.

For the rural sector and remote coastal areas and islands, small and community-size desalination and water purification units of appropriate capacity have good potential. Barge-mounted desalination units appear promising. Rural areas near large coastal cities or towns can be supplied with potable water from centralized large seawater desalination plants. Mobile trailer-mounted desalination units can also be taken from one place to another in case of natural disaster and emergency. Disposal of concentrate (brine)/reject water from the desalination plant is emerging as an environmental issue. Improper discharge has the potential to pollute the environment over a period of time. Different regions have different rules for disposing of the reject water. There are several options for safe disposal:

1. The feed is conditioned by pretreating it before it enters the RO modules. Dosing systems are used for pretreatment. This causes difficulties for disposal of the reject water. The dosing can be minimized using physical pretreatments such as MF, UF and NF processes.
2. The reject disposal problem can be addressed by adopting a combination of different membrane processes like NF and UF, or other hybrid processes. The possibility of removing toxic species can be explored using a combination of chemical processes with RO, NF and UF membrane processes. However, it may not be possible to solve the reject disposal problem entirely based on membrane process. Other alternatives must be explored.
3. Part of the reject water can be utilized for plantations that increase vegetation and decrease environmental pollution. Some species of halophytes such as *Atriplex* and *Salvadora* can grow even in high-salinity water and they make good cattle fodder.
4. Revenue can be generated from reject-water ponds by growing value-added materials. For example, algae like *Spirulina* can be cultured and grown in saline ponds.

Near coastal regions, excess withdrawal of groundwater may lead to an increase in groundwater salinity and disturb the fresh water table due to seawater intrusion. The drawing of groundwater for commercial purposes should be properly controlled, with alternate permanent water sources like seawater.

Cogeneration plants producing electricity and water appear advantageous as the cost is apportioned between the two major products: water and electricity. Facilities are shared. Different forms of energy such as thermal and electrical are available. Low-grade heat and waste heat are also available in significant quantity. Nuclear desalination is an attractive proposition in coastal areas, and nuclear energy has the potential to emerge as an important source of power. Desalination plants using nuclear energy (nuclear desalination) will come on stream as cogeneration projects. There is a need for more research effort and demonstration units in nuclear desalination, as well as the use of low-grade nuclear heat and waste heat for seawater desalination. Thermal desalination (MSF, MED) and hybrid technologies (MSF-RO, MED-RO) need to be further explored wherever excess energy is available, such as in refineries, steel plants and thermal power plants.

It is desirable to develop a data bank for bathymetric studies, detailed seawater analysis and oceanographic studies along the coast. A focused approach is required to achieve membrane development, better membrane performance, advanced pre-treatment, increased recovery, higher salt rejection, longer membrane life and better chemical resistance, while minimizing energy consumption and selecting proper construction materials.

The following areas of research and development, including demonstration, need to be further explored:

- Development of membranes having high permeate flux, long life, biodegradable characteristics, temperature-tolerant properties and biological fouling resistance.
- Establishing demonstration units using new energy and futuristic concepts.
- Simulation, design optimization, techno-economic analysis and environmental impact analysis.
- Databases containing information on raw water composition and its variations, with season and sensitivity analysis of desalination plant to such variation.
- Methodologies for up-scaling and down-scaling technologies.
- Advanced artificial intelligence (AI) integrated process control and instrumentation.
- Development of desalination plant management systems.
- Low-grade/waste heat utilization for water desalination.
- Hybrid systems.
- Recovery of strategic and precious materials from the brine rejected from the desalination plant.
- Advanced membrane-based pretreatment methods.
- Innovative intake and outfall structures.
- Innovative desalination processes.
- Cost-reduction strategies through technological innovations.
- Advanced heat transfer surfaces for thermal desalination systems.
- Energy consumption.
- Module configuration.
- Construction materials (such as concrete shell, plastic heat exchanger, etc).
- Environmental impact assessment of brine discharge from desalination plant.
- Setting up technology development mission projects in desalination resulting in a marketable product/patentable technology.
- Setting up desalination centers for research, training and technology demonstration at various scales, and promoting human resources.
- Studies on mixing patterns of brine in the ocean.

Some improvements in thermal processes could emerge in the future with the use of the NF process for pretreatment to control scaling at higher top brine temperatures,[2] and the development of thermal-based hybrid systems using renewable energies such as geothermal, solar or wind energy.[3–6]

9.6 INDUSTRY–ACADEMIA INTERACTION

It is important that industry (or a consortium of industries) collaborates with academia to jointly develop a product or process and share the intellectual property. Research-based assignments for exploratory and futuristic ideas have potential for new technological innovations. Academia–industry interactions are useful for solving an immediate problem, verifying a specific design, undertaking specialized testing, and designing and developing certain components or structures. Sponsored research needs to be focused on basic sciences or on applied areas where certain fundamental investigations are to be taken up. Sponsored research schemes are needed as long-term assignments in highly specialized fields for which applications are slowly developing.

Industry and academia together can identify projects that are of current or immediate relevance to industry. These may require continuous applied research and development in order to produce results of real benefit to society. Such projects need not be focused on a product but can also involve a general concept or an idea. The problems may not be suitable for normal consultancy and at the same time may be difficult to formulate as a sponsored project. An industry or a group of industries can jointly sponsor a project of interest to academia in an area where they agree to share the know-how generated. The project duration could be from six months to three years. In this approach, the emphasis may be on research to widen knowledge and to explore new avenues for development. Judicious planning can provide excellent benefits to the industries concerned. This could be linked with technology upgrades, transfer of know-how and international collaboration. Such studies could provide industry with insight into the new technologies that are likely to replace existing ones. Association with such projects is bound to offer specific benefits to industry in the long run.

Pilot plants may be developed involving collaboration between academia and industry. The pilot plant should eventually become self-sustaining based on consultancy projects from small and large-scale industries. 'Water technology centers of excellence' with technology innovation divisions could be established in different locations offering training, technical guidance and analytical support, and developing expertise.

Industry–academia association is one way of helping industries and academic institutions to get together and use analytical instruments, and laboratory, library and computer facilities. Industries with research funds might invest in desalination and water purification schemes to earn the goodwill of a particular local community by solving their water requirements.

9.7 BUSINESS MODELS

Desalination and water purification is a relatively costly proposition compared to natural water supply. However, with the growing deficit of fresh water, desalination and water recycling are the only way of supplementing existing resources. Industries by their very nature add value to the resources, hence any marginal cost addition can be easily absorbed without making any major impact on the overall economics. Even if the industries assume desalination cost as an additional cost, the increase in the overall cost of the ultimate product would only be marginal, mostly not exceeding 0.1% in most water-intensive industries, including power, textiles, paper, etc. Natural water without any contaminants such as F, As, NO_3, is a better option due to its natural taste compared to conditioned desalinated water. In view of this, it would be a good option to satisfy the potable water requirements using natural resources and to use industry to meet the additional demand on a full-cost basis. Such a policy decision would encourage industries to trim their water consumption, utilize their resources efficiently and above all set up their own desalination and water recycling plants. Several industries have found that water recovered from sewage has been cost-effective compared even to conventional water resources. Detailed surveys of the consumption patterns of different industries, the distribution network and socio-political factors, etc. may have to be carried out and factored in before an appropriate policy decision can be reached. Nevertheless the policy of using the appropriate quality of water for different uses may have to be cultured into society to ensure equitable distribution of resources.

Where desalination and water reuse plants are being run for industry, companies have to make a policy decision to either fund the capital cost and operate the plant themselves, or use the Design–Build–Own–Operate–Transfer (DBOOT) or Build–Own–Operate–Transfer (BOOT) concepts. Each method has its own merits. The former is advantageous as it provides flexibility in production patterns depending on demand. DBOOT/BOOT presupposes minimum production irrespective of other resources or fluctuations in consumption. For large plants meeting the needs of communities, DBOOT could be a good option despite the higher cost of water, as it ensures reliability. Normally, a reduction in demand is not expected in a large public supply system except perhaps during extraordinary monsoons. The case of small and medium desalination plants based on brackish water sources is complicated, as here production depends on logistics, infrastructure and local constraints including brine disposal and operator skills. In remote areas, desalination may be the last option. In order to ensure efficient operation and distribution, plants may have to be managed by local cooperatives.

9.8   CHALLENGES AND OPPORTUNITIES

It is time to realize that the value of water is dependent on its quality. As mentioned above, a dedicated water center of excellence funded by a consortium of government and industry could have a competitive edge as it would need to stay abreast of current and future trends in the water market. Such a center could interact with academia and industry to develop resource sharing and state-of-the-art components.

Alternative technologies should be explored, initially for technical feasibility and later with a focus on cost reduction, as the ultimate cost is a function of a complex mix of factors, both technological and environmental.

REFERENCES

1. Ghaffour, N., Missimer, T.M. and Amy, G.L. 2013. Technical review and evaluation of the economics of the water desalination: Current and future challenges for water supply sustainability. *Desalination* 309: 197–207.
2. Al-Sofi, M.A.K., Hassan, A.M., Mustafa, G.M., Dalvi, A.G.I. and Kither, M.N.M. 1998. Nanofiltration as a means of achieving higher TBT of more than 120°C. *Desalination* 118: 123–129.
3. Ghaffour, N., Reddy, V.K. and Abu-Arabi, M. 2011. Technology development and application of solar energy in desalination: MEDRC contribution. *Renewable and Sustainable Energy Reviews* 15: 4410–4415.
4. Mahmoudi, H., Saphis, N., Goosen, M., Sablani, S., Sabah, A., Ghaffour, N. and Drouiche, N. 2009. Assessment of wind energy to power solar brackish water greenhouse desalination units—A case study. *Renewable and Sustainable Energy Reviews* 13: 2149–2155.
5. Zejli, D., Ouammi, A., Sacile, R., Dagdougui, H. and Elmidaoui, A. 2011. An optimization model for a mechanical vapor compression desalination plant driven by a wind/PV hybrid system. *Applied Energy* 88: 4042–4054.
6. Gastli, A., Charabi, Y. and Zekri, S. 2010. GIS-based assessment of combined CSP electric power and seawater desalination plant for Duqum-Oman. *Renewable and Sustainable Energy Reviews* 14: 821–827.

10 Artificial Intelligence and the Internet of Things in Water Management

10.1 ARTIFICIAL INTELLIGENCE AND MACHINE LEARNING IN THE INTEGRATED WATER SECTOR

Artificial intelligence and the internet of things (AIoT) relates to how AI technologies and IoT infrastructure combine to achieve efficient operations and enhance human–machine interactions as well as data management and analytics. AI can be used to transform IoT data into useful information for better decision-making processes, creating a foundation for newer technology. AI and IoT are the two most disruptive types of technology in today's world, thanks to radical changes in computational power, the huge amount of data available and recent advances in deep neural networks. AIoT is translational and mutually beneficial as AI adds value to IoT through machine learning (ML) capabilities and IoT adds value to AI through connectivity and data exchange. ML enables operators to make informed decisions based on actionable insights. The predictive capabilities of ML are based on processing collected information, learning from data and providing possible outcomes. Integration of AI with IoT builds an intelligent network of connected things and people, enabling data collection from and communication with surrounding environments.

Although the applications of IoT-based AI-driven architecture are particularly relevant to the water sector, it can be extended to other areas, such as surveillance. As an advanced technology, AIoT application has potential to play an increasingly important role in water resource management, drinking water supply, wastewater treatment, real-time water monitoring and the smart water grid (Figure 10.1).

Predictive real-time monitoring of existing and future water purification units is an important area to be explored. AIoT with component life cycle prediction and leakage identification can play a meaningful role in the development of the smart water grid. AI/IoT-based online and real-time monitoring of water quality and flow rate in the drinking water pipeline network can contribute to efficient operation and provide consumers with greater confidence. AI/IoT-assisted process data trending gives real-time data on quantity and quality of drinking water supply. It helps to assess corrosion behavior of pipelines and enhance their service life as well as

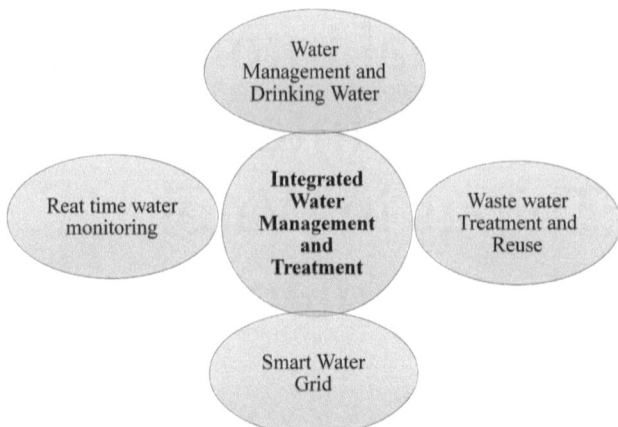

FIGURE 10.1   AIoT applications in water management.

minimize shutdown time. Predictive real-time AI/IoT-enabled systems also help in evaluating the performance of water purification units.

Currently sensors for online monitoring of some of the parameters, such as total dissolved salts (TDS) and pH, are available in the market, while sensors and associated systems for online monitoring of several other parameters need to be developed. This will help in the innovation ecosystem as well as catalyzing, mentoring and supporting best practices in technology translation and entrepreneurship. The possibilities are huge as there are several contaminants at any one time in water as well as new emerging contaminants. The world market is being flooded with IoT devices and AI skill sets are rapidly increasing.

10.2   AIOT-ENABLED WATER TREATMENT

The AIoT-enabled water treatment sector requires appropriately designed sensor modules and IoT devices along with AI techniques. The following AIoT components are important from a design and development point of view:

1. Algorithm design to bring intelligence component to IoT network.
2. Enabling local intelligence in the sensor modules and networks.
3. Development of online and query/event-based data acquisition system for monitoring.
4. Multimodal data analytics for analysis of data acquired from various types of sensors.
5. Developing combination of distributed and central decision-making architectures to control AIoT network.
6. Development of encryption/decryption algorithms, key distribution mechanisms, intrusion-detection systems and secure routing policies for sensor grid security and management.
7. Development of cyberphysical social system (CPSS).

AI-enabled sensor modules and arrays have different types of sensors. The nature of the signal at the output of the sensor can be physical, chemical or electrical. The sensor output is processed by the interface circuit which consists of a signal conditioning unit followed by data converters. A particular sensor requires a dedicated signal conditioning unit to make the sensor output fall in the desired range for converters. The purpose of data converters is to digitize the signal to make it compatible with displays or computers. The development of these sensors plays a crucial role in the AIoT-enabled water treatment system. Biosensors and chemical sensors are needed for water quality monitoring. AIoT and intelligent sensor modules make it possible to use precise amounts of the correct inputs.

10.3   AIOT-DRIVEN WATER MANAGEMENT

Fresh water is needed for different purposes such as agriculture, industry, drinking, cooking, etc. Agriculture needs large quantities of water for irrigation and good quality water for various production processes. Crops and livestock need water to grow. While feeding the world and producing a diverse range of non-food crops such as cotton, rubber and industrial oils in an increasingly productive way, agriculture is the biggest user of water. However, with the adverse impacts of climate change continuously increasing, conventional methods of resource-intensive farming may not be able to deliver sustainable food and agricultural production. Technological transformation of agriculture and water management systems is needed to meet future demands of per-capita food consumption. In many parts of the world, farmers often obtain sub-optimal yields from their crops due to inappropriate irrigation. Deployment of sensor arrays to measure parameters like pH, electrical conductivity (EC), Mg, K, organic matter content, moisture level, nitrogen level, etc., and creating an IoT network, have potential to enable field data collection and online water and crop monitoring. AIoT has the potential to make use of machine learning methods to analyze and process the collected sensor data to indirectly measure soil indicators and take preliminary decisions such as controlling the water supply. The agriculture sector is increasingly subjected to water risks because it consumes the largest share of total water resources for irrigation. Further, crops in different regions require different water quality. Improving water management in agriculture and water treatment are, therefore, essential for a sustainable and productive agro-food sector.

10.4   SENSOR DEVELOPMENT FOR AIOT IN THE WATER SECTOR

AIoT is directly linked to the area of sensor development for monitoring water quality. Current approaches to monitoring water quality include microbial and physico-chemical measurements.[1] The physico-chemical parameters are preferred as they can be analyzed more quickly than microbial parameters. The most common physico-chemical parameters include EC or TDS, pH, turbidity, oxidation reduction potential (ORP), temperature and chlorine content. These parameters can be monitored using different water quality sensors. Although current methodologies available in the literature perform thorough analysis, they lack real-time monitoring of these parameters to enable critical decisions in situations of poor water quality.[2]

Further, spatiotemporal coverage of the existing methodologies is poor as data samples are collected from only a few locations due to limited resources and labor.

Several miniaturized microbial fuel cells (MFCs) have been reported in the literature for monitoring of water quality and toxicants. For example, a microbiosensor electrode exhibited a linear relationship between 0 and 1.6 mM acetate concentrations with a 79 ± 8 µM limit of detection (S/N=2).[3] In another study, a paper-based MFC was fabricated for formaldehyde detection in water. The detection of formaldehyde was realized by a sudden drop in the current output of MFC.[4] A dried biofilm electrode prepared with non-pathogenic *Escherichia coli* culture was used in an air cathode MFC assembly. This micro-device could generate instant current upon exposure to a drop of tap water.[5, 6] Notwithstanding some of the studies mentioned above, the utility of pre-formed biofilms for MFC-based sensor applications needs to be explored, and their online performance needs assessment.

Measurements and collected data play a critical role in technological transformation. A major challenge in arid and semi-arid land is the type and quantity of water required as different crops require different water quality to grow. Similarly, the major challenge in piped water supply is to predict the component life cycle and identify leaks.

10.5   MAJOR FOCUS AREAS

Although there is a considerable amount of knowledge and literature available on both AI and IoT, their combination specifically for water management is yet to be established. As IoT networks become increasingly popular in industry, there is an increasingly large amount of unstructured machine data. The growing amount of human-oriented and machine-generated data drives substantial opportunities for AI support of unstructured data analytics solutions. Data generated from IoT-supported systems are extremely valuable, both for internal corporate needs as well as for consumer-facing functions such as product lifecycle management.

Water utilities equipped with IoT sensors and other data-driven technologies that constantly collect data during the different phases are able to draw information from those phases, thus helping to improve performance. AI and ML have the potential to create more effective water treatment processes, anticipate problem areas and help direct trouble-shooting efforts in those areas well ahead of time. ML can also provide new insights for planning the usage of water utilities as well as operation and maintenance expenditure and investment.

The application of AIoT enables technological development to address the challenges of ecological changes and their impact on water resources. Tools and technologies for area-specific applications such as grey water management and water resource management not only support the particular area but have an impact on the whole region.

Major focus areas include intelligent IoT networks, sensors with local intelligence, query/event-based data acquisition, multimodal data analysis, hybrid (distributed and central) decision-making architecture, sensor grid security and management, and CPSSs as well as smart water grids, real-time water monitoring, wastewater treatment and water reuse.

REFERENCES

1. Hall, J., Zaffiro, A.D., Marx, R.B., Kefauver, P.C., Krishnan, E.R., Haught, R.C., & Herrmann, J.G. 2007. On-line water quality parameters as indicators of distribution system contamination. *Journal of the American Water Works Association*, 99, 66–77.
2. Bhawan, P. 2005. *Water quality monitoring in India-achievements and constraints*, Article http://mdgs.un.org/unsd/environment/abst_wasess5a2india.doc
3. Lambrou, T., Anastasiou, C., Panayiotou, C., & Polycarpou, M. 2014. A low-cost sensor network for real-time monitoring and contamination detection in drinking water distribution systems. *IEEE Sensors Journal*, 14, 2765–2772.
4. Atci, E., Babauta, J.T., Sultana, S.T., & Beyenal, H. 2016. Microbiosensor for the detection of acetate in electrode-respiring biofilms. *Biosensors and Bioelectronics*, 81, 517–523.
5. Chouler, J., Cruz-Izquierdo, Á., Rengaraj, S., Scott, J.L., & Di Lorenzo, M. 2018. A screen-printed paper microbial fuel cell biosensor for detection of toxic compounds in water. *Biosensors and Bioelectronics*, 102, 49–56.
6. Nguyen, D.T., & Taguchi, K. 2019. A disposable water-activated paper-based MFC using dry *E. coli* biofilm. *Biochemical Engineering Journal*, 143, 161–168.

Index